Understanding the Universe

Understanding the Universe

The Physics of the Cosmos
from Quasars to Quarks

Andrew Norton
The Open University

CRC Press
Taylor & Francis Group
Boca Raton London New York

CRC Press is an imprint of the
Taylor & Francis Group, an **informa** business

First edition published 2021
by CRC Press
6000 Broken Sound Parkway NW, Suite 300, Boca Raton, FL 33487-2742

and by CRC Press
2 Park Square, Milton Park, Abingdon, Oxon, OX14 4RN

Library of Congress Cataloging-in-Publication Data

Names: Norton, Andrew J., author.
Title: Understanding the universe : the physics of the cosmos from quasars to quarks / Andrew John Norton, The Open University.
Description: First edition. | Boca Raton : CRC Press, 2021. | Includes bibliographical references and index.
Identifiers: LCCN 2020056848 (print) | LCCN 2020056849 (ebook) | ISBN 9780367748050 (hardback) | ISBN 9781003164661 (ebook)
Subjects: LCSH: Cosmology. | Particles (Nuclear physics)
Classification: LCC QB981 .N775 2021 (print) | LCC QB981 (ebook) | DDC 523.1--dc23 LC record available at https://lccn.loc.gov/2020056848LC ebook record available at https://lccn.loc.gov/2020056849

ISBN: 9780367748050 (hbk)
ISBN: 9780367759322 (pbk)
ISBN: 9781003164661 (ebk)

Typeset in Times
by Deanta Global Publishing Services, Chennai, India

Dedication

For Joseph John Norton (1932–2020)

Contents

PART II   The Large-Scale Universe

PART III   Universal Processes

Author

Andrew Norton is Professor of Astrophysics Education in the School of Physical Sciences at the Open University and is a former vice president of the Royal Astronomical Society. He earned his PhD in X-ray astronomy from Leicester University working on interacting compact binary stars. His current research focusses on time domain astrophysics from large-scale photometric surveys, including variable stars and transiting exoplanets. In his role as an educator at the Open University, he has taught many areas of physics, exoplanetary science, stellar astrophysics, accretion physics, theoretical and observational cosmology, extragalactic astrophysics, and practical observational astronomy using optical and radio telescopes. He has been academic consultant for several OU/BBC TV co-productions and was co-author of the OU's "60 second adventures in astronomy" videos. He has an Erdös-Bacon-Sabbath number of 13.

1 Quasars to Quarks

This book's title is very ambitious, but the intention is indeed to show how physics allows us to understand the way in which the Universe operates, from the very large scale of quasars to the very small scale of quarks, and everything in between. Almost by definition, the rules of physics that have been discovered over the last 400 years or so allow us to understand the Universe and so explain how it works on all scales. In order to explore the Universe, I will therefore tell the story of its entire history, from its origin 14 billion years ago to the present day, and examine both the smallest objects and the most distant structures that comprise it. Along the way, many of the fundamental principles of physics will be encountered, and by the end of the book you will have met the key ideas of two specific areas of science, namely cosmology and particle physics.

To set the scene for this incredible journey it's convenient to begin by looking at the various sizes of things, and the various distances to objects, that comprise the Universe. In scientific terms, these sizes and distances are jointly referred to as length scales (Figure 1.1). The journey from one extreme of length scale to the other might be summed up by the phrase "quasars to quarks." Although the words "quasar" and "quark" are adjacent to each other in the index to this book, the objects they refer to are as far apart as it's possible to be in terms of the length scales of the Universe. Quasars represent the most distant astronomical objects it is possible to observe and are up to one hundred million billion billion metres away. This distance may be written as 100 000 000 000 000 000 000 000 000 m or more compactly as 10^{26} m, where the positive superscript value of "26" represents the "number of tens" that must be multiplied together to get the specified value. In contrast, quarks are the fundamental constituents of matter and are smaller than one-tenth of a billionth of a billionth of a metre in size. This size may be written as 0.000 000 000 000 000 000 1 m or more compactly as 10^{-19} m, where this time the *negative* superscript value of "–19" indicates the "number of tens" multiplied together that must be divided into "1" to get the specified value. The distance to a quasar is therefore 45 factors of 10 greater than the size of a quark; in other words it is 45 orders of magnitude larger. These two length scales are separated by a factor of a billion, billion, billion, billion, billion – conveniently represent the extremes of human comprehension of the Universe.

In science, prefixes to indicate multiples of units are often used to avoid having to write such large or small numbers. Ones you may be already familiar with include nano- (n, 10^{-9}), micro- (μ, 10^{-6}), milli- (m, 10^{-3}), kilo- (k, 10^{3}), mega- (M, 10^{6}), and giga- (G, 10^{9}). You may have seen these in units such as nanometres (nm), microseconds (μs), millilitres (ml), kilograms (kg), megayears (My), and gigajoules (GJ), for instance.

FIGURE 1.1   Length scales in the Universe. Quarks (10^{-19} m) are a billion times smaller than atoms (10^{-10} m), which are a billion times smaller than apples (10^{-1} m), which are themselves a billion times smaller than Jupiter (10^8 m). The distances to the nearest stars (10^{17} m) are a billion times larger than the size of Jupiter, and quasars are a billion times further away (10^{26} m) than nearby stars.

   Before beginning our exploration of the Universe, it's worth pausing for a moment to try to appreciate the sheer range of length scales implied by the simple phrase "quasars to quarks." A factor of a billion (10^9) spans nine orders of magnitude. In terms of sizes that we can more easily comprehend, this is equivalent to going from, say, the size of the planet Earth (to the nearest order of magnitude this is 10,000 km, or 10^7 m, in diameter) down to the size of a marble (about 10 mm, or 10^{-2} m, in diameter). But there are *five* such steps, *each* requiring a decrease in size by a factor of a billion, in going from quasars to quarks!

Understanding the Universe throughout this range of length scales is necessarily the most wide-ranging subject that can be addressed by science. Other scientific issues, such as climate change or genetic engineering, certainly have more immediate relevance to everyday lives, but when it comes to fundamental questions such as:

- How does the Universe behave on small and large scales?
- What rules does the Universe follow?
- How does the Universe change with time?

there are none that are larger in scope. Answers to questions like these are to be found in the fields of cosmology and particle physics. Scientists who work in these two apparently unrelated areas of science – one concerned with the infinitely large, the other with the unimaginably small – now work together in an attempt to explain the universal processes that occur throughout time and space.

1.1 COSMOLOGY AND PARTICLE PHYSICS

In this book I will explore the physics of both the very large and the very small to take you on a journey from quasars to quarks. Along the way, many other components of the Universe will be encountered, and I will introduce you to the key ideas of modern physics, developed over the last century, that now form the fundamental basis for understanding the Universe.

Cosmology is the branch of science that involves the study of the Universe as a whole. The research tools of cosmologists include powerful telescopes that can detect galaxies out to the furthest reaches of the Universe (Figure 1.2). It may seem strange that people working in this field should count those who work on particle physics amongst their closest allies. The research tools of particle physicists include giant particle accelerators in which high-energy beams of subatomic particles are smashed together, enabling details of exotic reactions to be investigated and understood. But

(a) (b)

FIGURE 1.2 (a) Telescopes on Mauna Kea, Hawaii and (b) part of the Large Hadron Collider particle accelerator at CERN, on the Swiss/French border.

this is the key to the union of these two subjects. For only in particle accelerators are scientists able to recreate the high-energy conditions that once existed in the Universe during the first moments of its creation. When particle physicists study these reactions, they can provide cosmologists with a window on the Universe when it was only one-thousandth of a billionth of a second old.

An example of the interplay between these two areas of study concerns the fundamental particles known as neutrinos, which will feature later in the book. Some years ago, cosmologists studying reactions that occurred in the early Universe announced that there can be no more than three types of neutrino. If there were, say, four types of neutrino then they calculated that there would be more helium in the Universe than is actually observed. Particle physicists, studying decays of exotic particles in high-energy accelerators, were also able to calculate how many types of neutrino there are in the Universe. The answer the particle physicists arrived at was also three – if there were more, or fewer, types of neutrino, the particles they were studying would have decayed at a different rate. So, it is highly likely that there really are only three types of neutrino in the Universe – whether the problem is tackled from the large or the small scale.

1.2   UNDERSTANDING HOW THE UNIVERSE WORKS

The big questions outlined at the start of Chapter 1 will now be used as themes for what follows. To begin with, Chapter 2 summarizes some of the key ideas that form the foundations of physics. This includes information about atoms, motion, energy, and light that are central to understanding the world around us. Material in this chapter is your toolkit for understanding what follows and will highlight the importance of these tools for understanding the Universe.

In the first part of the book, comprising Chapters 3 to 7, the topics are the overall structure and composition of matter on the smallest scales. The branch of science that describes how the smallest particles of matter behave is known as quantum physics. To begin with, the two key features of quantum physics that control the behaviour of matter and energy at a fundamental level – quantized energy and indeterminate positions and velocities – are explored. Taking the world apart, you'll examine the structure of atoms before moving on to look at how atomic nuclei behave, and finally you will look inside the protons and neutrons to discover quarks – the fundamental building blocks of the Universe. This is quite an itinerary for your journey into the heart of matter. You will be working in the tiny, subatomic domain and will be coming to terms with ideas that are almost unbelievable in the context of the everyday world. Most people are awe-struck by spectacles such as mountain scenery here on Earth or an exploding star in outer space. Prepare yourself to encounter phenomena that are much too small to be seen but that are none the less just as rich in fascination and mystery.

The second part of the book, comprising Chapters 8 to 10, turns to the science of cosmology and examines the overall properties of the Universe on the largest scales. After exploring just *how* astronomers observe the Universe using telescopes that operate across the electromagnetic spectrum, you will see that observations of quasars hold the key to understanding the distant Universe – by which is meant distant in both time and space. You will discover that the Universe is not static: it was different in the past to how it is now, and it will be different again in the future. Understanding this evolution relies crucially on two pieces of evidence: first, evidence that the Universe is *expanding* and, second, evidence that the Universe is *cooling*. Each of these is considered in turn to complete the picture of the Universe's large-scale behaviour. Once again, you will be dealing with concepts that lie completely outside those of everyday experience.

Any attempt to chart the history of the evolving Universe must take account of the laws that govern all physical processes. So, the third part of the book, comprising Chapters 11 to 15, contains accounts of the distinctive features of the four types of interaction of matter and radiation: electromagnetic, strong, weak, and gravitational forces. You will see that these four interactions underlie *all* processes, at *all* scales, *everywhere* in the Universe. Then the question is raised of whether the four interactions are actually distinct or whether there might be bigger and better theories, that unify some of the four interactions.

Bringing together the information from the earlier chapters, the final part of the book, comprising Chapters 16 to 18, begins by presenting a history of the Universe from the Big Bang to the present day. It then explores the galaxies, stars, and planets that are observed in the Universe today, and finally considers what the future of the Universe may have in store.

Some of the ideas discussed in this book may challenge the view of the world that you currently hold, and throughout history such challenges have been one of the hallmarks of scientific progress. Apart from the intellectual excitement of topics in cosmology and particle physics, they also serve to illustrate the way in which scientists continually strive to push back the boundaries of knowledge, extrapolating from what can be measured in the laboratory to realms that are impossible to study directly. However, it's important to note that these subjects are, nonetheless, relevant to everyday lives – for instance, the current technological age could not have come about were it not for the underpinning science of quantum physics. You will also encounter some rather bizarre ideas in the following pages. These will include particles that appear out of nothing, gravitational waves that permeate the entire Universe, and a remarkable theory of an 11-dimensional spacetime! Prepare for some mental exercise as you embark on a journey to the frontiers of physics and an exploration of the processes that allow us to understand the Universe.

2 The Physical World

Physics is the science that provides the rules for understanding the Universe, and many of the underlying ideas of physics are probably familiar to most people, whether they realize it or not. This chapter will remind you about (or introduce you to) the concepts of atoms, motion, energy, and light that are fundamental to understanding what follows later in the book.

2.1 THE WORLD OF ATOMS

Everything on Earth is composed of atoms, and there are known to be around 90 different types of atom that occur naturally in the world. A material made of a single type of atom is known as an element. The most abundant elements in the Universe are those composed of the two simplest atoms: hydrogen and helium. Later in this book, you will see that these elements emerged from the Big Bang in the first few minutes after the Universe began. Here on Earth, there are also significant amounts of other elements, in particular: carbon, nitrogen, oxygen, sodium, magnesium, aluminium, silicon, sulphur, calcium, and iron. Later on, you will see that these elements emerged from processes occurring within stars that lived and died long before the Earth was formed.

2.1.1 ATOMIC COMPONENTS

Whatever the type of atom, each one has certain features in common. Each contains a central nucleus, which carries a positive electric charge as well as most of the atom's mass. The nucleus is surrounded by one or more negatively charged electrons (symbol e⁻); they each have a much smaller mass than the nucleus. The nucleus of an atom is what determines the type of element. The very simplest atoms of all, those of the element hydrogen, have a nucleus consisting of just a single proton (symbol p). The next simplest atom, helium, has two protons in its nucleus; lithium has three protons; beryllium has four; boron has five; carbon has six; and so on, all the way up to uranium with 92 protons. The *number* of protons in the nucleus of an atom is known as its atomic number (represented by the symbol Z). As you know, elements have very different chemical properties. Later in this book, you will see why this is so.

The electric charge of a proton has a numerical value of about $+1.6 \times 10^{-19}$ coulombs (where the unit of electric charge has the symbol C). The electric charge of an electron is exactly the same as that of a proton, but negative instead of positive, so it has a value of about -1.6×10^{-19} C.

The other constituents of atomic nuclei are neutrons (symbol n); they have a similar mass to protons but have zero electric charge. Normal hydrogen atoms have no neutrons in their nuclei, although there is a form of hydrogen – known as deuterium – that does. The nucleus of a deuterium atom consists of a proton *and* a neutron. It is still the element hydrogen (because it contains one proton) but it is a "heavy" form of hydrogen, thanks to the extra neutron. Deuterium is said to be an isotope of hydrogen. Similarly, normal helium atoms contain two neutrons in their nucleus, along with the two protons; but a "light" isotope of helium, known as helium-3, contains only one neutron instead (Figure 2.1). The combined number of protons and neutrons in the nucleus of an atom is the mass number (represented by the symbol A) of the atom. Isotopes therefore denote forms of the same element with different mass numbers. The nucleus of a particular isotope is referred to as a nuclide.

As a short-hand, isotopes of each atomic element may be represented by a symbol. Letters are used to indicate the name of the element itself, and two numbers are used to indicate the atomic number (lower) and mass number (upper). So a normal hydrogen atom is represented by ^1_1H, and an atom of the heavier isotope, deuterium, by ^2_1H; regular helium is ^4_2He and light helium is ^3_2He. Other familiar atoms mentioned earlier have symbols as follows: carbon (C), nitrogen (N), oxygen (O), sodium (Na), magnesium (Mg), aluminium (Al), silicon (Si), sulphur (S), calcium (Ca), and iron (Fe).

Sometimes, protons and neutrons are collectively referred to as nucleons since both types of particles are found inside the nucleus of an atom. Similarly, electrons, protons, and neutrons are often collectively referred to as subatomic particles, for obvious reasons.

Normal atoms are electrically neutral, so the positive electric charge of the nucleus is exactly balanced by the negative electric charge of the electrons surrounding it. Because each electron carries an electric charge that is equal and opposite to that of each proton, the number of electrons in a neutral atom is *exactly* the same as the number of protons in its nucleus. If one or more electrons are removed from,

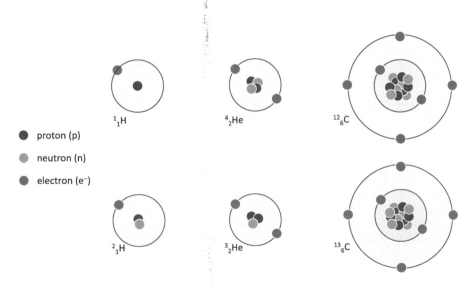

FIGURE 2.1   Atoms and isotopes of hydrogen, helium, and carbon.

or added to, a neutral atom, then an electrically charged ion is formed. Positive ions have had electrons removed from the neutral atom, whilst negative ions have had electrons added to the neutral atom.

Chemists find it convenient to think of the electrons in an atom as being arranged in a series of concentric shells and subshells surrounding the nucleus. While this description helps explain why certain atoms interact with others in particular ways, it is a model that only partially explains the behaviour of atoms. You might be asking yourself, *why* do electrons arrange themselves in this way? In the following chapters you will see how the shell picture emerges.

Finally, consider the size of atoms and nuclei. Whereas a typical atomic nucleus has a size of around one-hundredth of a billionth of a millimetre (10^{-14} m), the size of the atom itself is determined by the size of the region occupied by the electrons that surround the nucleus. The overall size of an atom is ten thousand times larger, namely about one-tenth of a millionth of a millimetre (10^{-10} m) across. For comparison, if an atom were scaled up to a size of 40 km in diameter (that's about the size of the M25 motorway around London), then the nucleus would be only 4 m across (which is about the length of a taxi in the centre of the city). Another analogy that you may prefer to visualize is that of a flea perched on the centre spot of the pitch at Wembley Stadium, compared to the size of the entire stadium itself. Either way, the best picture of an atom is one of mostly empty space but containing a tiny central core (the nucleus) surrounded by a vast cloud of electrons.

You may be wondering, if atoms are mostly empty space, why do they not routinely pass through each other? When an apple is placed on a table, what's to stop the atoms comprising the apple from passing easily through the atoms comprising the table, making use of all that empty space to simply slide past each other? The answer is that the apple and the table are held apart by the electrical force of repulsion between the electrons in their constituent atoms. You will read more about such electrical forces in Chapter 11.

2.1.2 MOLECULES AND STATES OF MATTER

Atoms may combine with each other, and with atoms of different kinds, to form molecules. Small molecules may contain just a few atoms. Examples found in various astronomical contexts include carbon monoxide (CO), carbon dioxide (CO_2), water (H_2O), ozone (O_3), ammonia (NH_3), and methane (CH_4). In each case, the subscript indicates the number of atoms of a given type within the molecule; if no number is shown it is implied that only one atom of the specified type is present in the molecule. Larger molecules also exist, with more complex chemical formulae. In astrophysics, many such molecules have been found in interstellar space, including ethanol (CH_3CH_2OH), urea (NH_2NH_2CO), benzene (C_6H_6), and buckminsterfullerene (C_{60}) to name just four examples.

When atoms and molecules aggregate in large numbers, substances can take one of several physical states, usually described as solids, liquids, and gases. In a solid, such as graphite or diamond composed of atoms of carbon, or water ice composed of water molecules, the atoms or molecules are locked into a rigid lattice structure. They oscillate around fixed equilibrium positions, with the amount of vibration increasing as the temperature rises. Conversely, in a gas, the atoms or molecules are free to move individually and follow zigzag, highly erratic paths. Liquids are an intermediate case. Here the molecules oscillate for a while in the local environment of their neighbours, but then move on to join another group of molecules. In astronomy, we most often deal with substances in a gaseous state, although solids such as ices and silicate minerals are present in planets and planet-forming environments too.

Gases have the most straightforward behaviour and may usefully be characterized in terms of their pressure, volume, and temperature. Some simple relationships are observed as follows. For a fixed amount of gas contained in a fixed volume (such as a sealed rigid container), increasing its temperature will cause the pressure to rise. Similarly, if a fixed amount of gas expands at constant temperature, its pressure will fall. Finally, if a fixed amount of gas at constant pressure is heated up, it will expand. Observations such as these may be familiar to you from everyday experience; the same laws hold throughout the Universe. Understanding the behaviour of gases is vital for figuring out how stars behave as they form and evolve, as you will see later in the book.

2.2 THE WORLD IN MOTION

The physical world may be understood as having three spatial dimensions plus a fourth dimension of time. The idea that objects move through space as time progresses is fundamental to the view of the world around us. Physics enables this motion to be described and quantified, and it also allows predictions to be made of future behaviour, based on what is measured at the current time. To understand the Universe, it is useful therefore to understand motion.

As a taste of what's to come, an idea that recurs throughout physics is that of conservation laws. You will read about three of these – the conservation of linear momentum, the conservation of angular momentum, and the conservation of energy – in the following sections. In 1918, the German mathematician Emmy Noether demonstrated that conservation laws in physics each result from an associated symmetry of a physical system, and this has become known as Noether's theorem. In particular, she showed that the three conservation laws mentioned above result from the fact that the laws of physics do not change with position, with time, or as the system is rotated.

2.2.1 DESCRIBING MOTION

Virtually everything in the Universe is in a constant state of motion – moons orbiting planets, stars moving within galaxies, and galaxies interacting with each other, to give just three examples. Knowing how to describe motion is therefore vital to understand the Universe.

A key idea in describing motion is that of the position of an object, which may be specified by three spatial coordinates. Near the surface of the Earth, latitude, longitude, and altitude might be used for these coordinates. In general terms a set of axes at right angles to each other may be constructed (often labelled as x, y, z), with a position labelled by its distance along each axis. In fact, various coordinate systems may be used, but since space is three-dimensional, three coordinates are always necessary unless the situation involves a restricted description such as motion on a surface (two dimensions) or motion along a line (one dimension). The displacement of an object is its distance from the origin of the coordinate system, in a specified direction.

The velocity of a moving object is the rate of change of its position with respect to time, and so has the standard unit of metres per second (written as m s^{-1}).

Considering motion in one dimension, on a graph of position against time, the velocity would simply be given by the gradient of the graph. If the velocity of an object is constant, the position–time graph would be a straight line; non-uniform motion would be characterized by a position–time graph that is a curved line. In such circumstances the rate of change of position with respect to time may vary from moment to moment and so define the instantaneous velocity. Its value at any particular time is determined by the gradient of the tangent to the position–time graph at that instant.

The instantaneous acceleration of a moving object is the rate of change of the instantaneous velocity with respect to time, and so has the standard unit of metres per second per second (written as m s^{-2}). Its value at any time is determined by the gradient of the tangent to the velocity–time graph at that instant.

If motion occurs in two or three dimensions, then we may conveniently use components of the displacement, velocity, and acceleration in each dimension. If the two or three components are mutually at right angles to each other, then their net result may be obtained using Pythagoras's theorem which states that the square of the magnitude of the resultant is equal to the sum of the squares of the magnitudes of the components.

Displacement, velocity, and acceleration are all quantities that have a magnitude *and* an associated direction. As such they are said to be vector quantities. The magnitude of each of these quantities is a positive number that carries no directional information. The magnitude of the displacement of one point from another represents the distance between those two points, and the magnitude of an object's velocity represents its speed. Quantities represented by a magnitude only are said to be scalar quantities. Since velocity comprises a speed *and* a direction, a change in either the speed *or* the direction of an object constitutes an acceleration.

A final quantity used to characterize motion in a straight line is that of linear momentum, sometimes called translational momentum (or just momentum for short). This is equal to the mass of an object multiplied by its velocity, so it too has both a magnitude and an associated direction, the same direction as that of the object's velocity. As with velocity, up to three mutually perpendicular components of linear momentum may be specified if necessary. The standard unit for momentum is kilogram metres per second (written as kg m s^{-1}); it quantifies an object's resistance to change by an external force – an object with a larger momentum is harder to slow down (or speed up) than an object with a smaller momentum. The momentum of a body, or system of bodies, is *conserved* unless an external force acts on the body. Noether showed that the conservation of linear momentum is simply a consequence of the fact that the laws of physics do not change from point to point throughout space.

2.2.2 MOTION IN A CIRCLE

A study of periodic motion is particularly useful in studying the Universe, where bodies are often in orbit around other bodies: planets around stars, stars around each other, and groups of stars around the centre of galaxies, for instance. By its definition, periodic motion is repetitive, and the simplest such motion to consider is that of motion in a circle.

You are probably familiar with the idea that a circle contains 360 degrees of angular measure. Perhaps slightly less familiar is that a circle also constitutes 2π radians. The Greek letter π (written as "pi" and pronounced the same as the word "pie") represents the ratio of the circumference of a circle to its diameter. This ratio is constant for *all* circles and has a value of about 3.14. So there are about 6.28 radians in a circle and 1 radian is equivalent to just under 60 degrees. An arc of a circle with a length equal to the circle's radius is therefore subtended by an angle of 1 radian at the centre of the circle.

The angular speed of motion is the rate of change of the angle between an object's position and the axis of measurement, and so is measured in units of radians per second. A particle in uniform circular motion completes one revolution (2π radians or 360 degrees) in one period. The instantaneous velocity is tangential to the circle and has a magnitude given by the angular speed multiplied by the radius of the circle.

The centripetal acceleration is directed towards the centre of the circle and has a magnitude given by the angular speed squared multiplied by the radius. In the situations encountered in astrophysics, the force responsible for providing the centripetal acceleration of an orbiting body is that of gravity, about which you'll learn more later.

Finally, the counterpart to linear momentum for motion in a circle is known as angular momentum. For a body in orbit around another, the magnitude of the angular momentum is given by its mass multiplied by its angular speed and the radius of the circle squared. In a similar way to linear momentum, the angular momentum of a body is conserved unless it is acted on by an external torque, a concept which is described below. Noether showed that the conservation of angular momentum is simply a consequence of the fact that the physics of a system is unchanged if it is rotated by any angle about an axis.

2.2.3 NEWTON'S LAWS OF MOTION

The three laws of motion that were derived by Isaac Newton in the 17th century lie at the heart of predicting how bodies will move. He published them in his book *Mathematical Principles of Natural Philosophy* in 1687.

According to Newton's first law of motion, a body remains at rest or in a state of uniform motion unless it is acted on by an unbalanced force. An "unbalanced force" is simply one that has a net result which is not zero. Clearly, pushing on an object with a force from the left whilst simultaneously pushing on the object with another force from the right which is equal in magnitude but opposite in direction to the first will produce no change to the object's state of motion – the net force is zero and there is no unbalanced force acting. The law therefore introduces force as a quantity that changes the motion of an object by causing it to accelerate. Like velocity and acceleration, any force has both a magnitude and a direction associated with it. It was noted above that changing an object's speed *or* direction constitutes an acceleration, so if an unbalanced force acts on a body then, according to Newton's first law of motion, either the speed of the body *or* its direction of motion will change (or both may change).

The description of a body's motion depends on the frame of reference from which the motion is observed. A frame of reference is a system for assigning coordinates to an object's position and times to events that occur. An inertial frame is a frame of reference in which Newton's first law holds true; it is a frame of reference that is not itself accelerating. Any frame that moves with constant velocity relative to an inertial frame, while maintaining a fixed orientation, will also be an inertial frame.

According to Newton's second law of motion: an unbalanced force acting on a body of fixed mass will cause that body to accelerate in the direction of the unbalanced force. The magnitude of the force is equal to the product of the mass and the magnitude of the acceleration. Force is measured in the standard unit of newtons (symbol, N), where 1 newton is the force required to accelerate a mass of 1 kilogram at a rate of 1 metre per second per second. A concept related to that of force is pressure, which is simply the force acting per unit area, and so is measured in the unit of $N\ m^{-2}$ or pascals (symbol, Pa).

The concept of momentum mentioned earlier leads to an alternative expression of Newton's second law of motion, namely that the total force acting on an object is equal to the rate of change of (linear) momentum of the object. Conversely, if no net force acts on a body then its momentum will not change, so momentum is conserved if no net force is acting.

For rotational motion, the equivalent idea to a force is that of a torque or turning force. When describing rotational motion, therefore, Newton's second law may be stated as saying that the total torque acting on an object is equal to the rate of change of angular momentum of the object. Conversely, if no net torque acts on a body then its angular momentum will not change, so angular momentum is conserved if no net torque is acting.

Finally, according to Newton's third law of motion, if body A exerts a force on body B, then body B also exerts a force on body A. These two forces are equal in magnitude but point in opposite directions. The law therefore indicates that "to every *action* there is an oppositely directed *reaction* of equal magnitude," as it is often expressed colloquially. An everyday example of this law in action is seen if you place an apple on a table. The apple (with a mass of about 100 g) exerts a downward force (due to its weight) on the table of around 1 newton and the table simultaneously exerts an upward (reaction) force on the apple which is exactly equal in magnitude to the weight, but acting in the opposite direction. There is no unbalanced force acting so the apple remains stationary.

Observers who attempt to apply Newton's laws in a non-inertial frame of reference (i.e. one which is itself accelerating) will observe phenomena that indicate the existence of fictitious forces, such as centrifugal force and Coriolis force. Centrifugal force is what makes you lean to the left as the car you're in turns to the right; Coriolis force is what gives rise to cyclones in weather systems on the rotating Earth. These phenomena are real, but the fictitious forces are not; they appear because of the acceleration of the observer's frame of reference relative to an inertial frame. By contrast, centripetal force is a real force in the sense that it arises in an inertial frame of reference. If you swing a bucket of water around your head on the end of a length of string, the tension in the string is due to centripetal force (although it's the centrifugal force that stops the water falling out of the bucket).

2.2.4 Relativistic Motion

The mostly intuitive ideas about describing motion embodied in Newton's laws turn out *not* to be true when the speeds involved approach the speed of light. In attempting to understand the Universe, objects moving with such relativistic speeds, as they are known, are commonplace. In this circumstance, the formulae derived under Newton's assumptions turn out to be only approximations to the real situation. For a more accurate description of nature it is necessary to turn to the theory of special relativity derived by Albert Einstein and published in 1905. Einstein's theory is based on two principles:

1. *The principle of relativity*: the laws of physics can be written in the same form in all inertial frames.
2. *The principle of the constancy of the speed of light*: the speed of light in a vacuum has the same constant value in all inertial frames, about three hundred thousand kilometres per second (or 3.00×10^5 km s^{-1}).

These two simple ideas lead to many intriguing and counter-intuitive results. Amongst these is the fact that the duration of a time interval is a relative quantity: the rate at which a clock ticks depends on the frame of reference in which it is measured. This is often paraphrased as "moving clocks run slow" and referred to as time dilation. Another result is that length is also a relative quantity. The length of a rod depends on the frame of reference in which it is measured. This is often paraphrased as "moving rods contract in the direction of motion" and referred to as length contraction. These effects are real but are simply not noticeable at everyday speeds; they only become apparent when the speeds concerned approach the speed of light.

An example of time dilation and length contraction in action concerns the subatomic particles known as muons. As you'll see in Chapter 7, these are a somewhat heavier counterpart of the electron which decay into electrons with a typical lifetime of around 2 microseconds. A ready source of muons is from cosmic rays (high-energy nuclei originating from the depths of space) that strike atoms in the Earth's upper atmosphere. If the number of muons liberated in the upper atmosphere is measured at the top of a mountain, then knowing the lifetime of a muon and the time that elapses before they reach the ground, it's possible to calculate the proportion of muons that should survive to be measured at sea level. However, far more muons survive to reach sea level than would be predicted based on non-relativistic assumptions. This can be understood by the realization that, for the rapidly moving muons, time runs more slowly, so allowing more of them to survive the trip. Alternatively, this may be understood by realizing that length is also contracted for the muons. So, the distance they travel from the top of the mountain to sea level is shorter than that which a stationary observer would measure, so once again allowing more to survive the journey. These predictions of special relativity are measured and confirmed to a high level of accuracy on a routine basis in laboratories around the world. It's not just that time "seems" to pass at a different rate, or objects "seem" to become shorter – these are *real* physical effects.

To appreciate the effects of special relativity, consider the following thought experiment. Imagine a spaceship setting off for the nearby star system, Epsilon Eridani, which is about 10 light-years away, travelling at 99% of the speed of light. Onboard is one of a pair of twins, whose sibling remains at home. Once the traveller reaches her destination, she sends a radio signal back to Earth, reporting that she's arrived. On Earth, that signal would arrive 20.1 years after the spaceship set off. Knowing that the radio signal took 10 years to get to Earth, the stay-at-home twin would deduce that it took the spaceship 10.1 years to make the trip (as expected, since that's just 10 light-years ÷ 0.99 × the speed of light). But for the travelling twin, the journey to Epsilon Eridani would only have taken 1.4 years, according to any clock she carried with her and according to her own body clock!

If, as soon as she arrived at Epsilon Eridani, the traveller then turned her spaceship around and immediately set off back to Earth, again travelling at 99% of the speed of light, she would arrive home 20.2 years after she left. That's just 10.1 years for each leg of the journey, as measured on Earth. The Earth-bound twin would be 20.2 years older than at the start of the adventure, but the travelling twin would only be 2.8 years older than when she had set off. This shows that time travel into the future is a very real possibility when special relativity is taken into account.

In everyday situations it is a familiar idea that relative speeds may be combined in a simple manner. For example, two cars travelling towards each other along a straight road, each moving at 100 km h^{-1}, will have a relative speed of approach of 200 km h^{-1}. Similarly, if one car travelling at 120 km h^{-1} overtakes another that is moving at 100 km h^{-1}, the relative speed between them will be just 20 km h^{-1}. This is in accord with what's known as Galilean relativity, outlined by the Italian scientist Galileo Galilei in 1632. Extending this idea to consider beams of light, think about the speed at which the photons emitted by the car's headlights and tail-lights travel. Imagine that you are travelling at 100 km h^{-1} in one of the cars mentioned above. You might think that the photons from the headlights of the car speeding towards you at 100 km h^{-1} would be observed to travel at the speed of light *plus* 200 km h^{-1}, or that the photons from the tail-lights of a car overtaking you at 120 km h^{-1} would be observed to travel at the speed of light *minus* 20 km h^{-1}. In fact, Einstein's second principle, stated above, says that this is *not* the case – the speed of light has precisely the *same* value no matter what the relative speed of the emitter or receiver of the light. In general, relative speeds of any objects do not simply combine by adding or subtracting, but the deviation from this simple rule only becomes apparent when dealing with speeds that approach the speed of light, which is seen to be a universal maximum limit.

One of the most dramatic consequences of the relative nature of time is that the order in which two events occur can, in certain circumstances, depend on the frame of reference of the observer. However, the *cause* of a particular event must always be observed to precede its *effect*. This causality will be preserved as long as the speed of light is the maximum speed at which a signal can travel.

2.3   THE WORLD OF ENERGY

If there is one equation in physics that many people have heard of, it's Einstein's famous statement: $E = mc^2$. What Einstein means by this equation is that energy (E) and mass (m) are interchangeable. Energy may be converted into mass and mass may be converted into energy. The conversion factor linking the two is the speed of light (c) squared. It turns out that the interconversion between matter and energy is vital for understanding the Universe, both in its earliest moments and in other energetic environments that are seen today.

Mass is a physical property that quantifies the amount of matter in a body. If, for simplicity, you consider an object like a diamond that is composed entirely of carbon atoms, then a more-massive diamond will contain more atoms of carbon than a less-massive diamond does. (Be careful not to confuse mass with weight. The weight of an object is a measure of the force of gravity acting on the object, and so may differ depending on where the object is located.)

The conservation of energy is a fundamental principle of the Universe, and one definition of energy is simply that it is the "stuff" that is conserved during any physical process. Alternatively, energy can be said to be a physical property possessed by an object that measures its capacity to make changes to other objects. There are a variety of possible changes, and these include changes in speed of motion, changes in temperature and changes in position with respect to other massive or electrically charged objects. Noether showed that the conservation of energy is simply a consequence of the fact that the laws of physics do not change with time.

Perhaps the most familiar form of energy is kinetic energy, or energy of movement. In everyday situations, the kinetic energy of an object is quantified by the expression: one-half of its mass multiplied by its speed squared. Some other types of energy fall into the category of potential energy, namely energy that is stored and which depends on the position of an object. Gravitational energy and electrical energy are two forms of potential energy; electromagnetic radiation, such as light, also carries energy.

The standard unit of energy is the joule (symbol J), and an electric kettle uses about 1000 joules of energy per second when heating up water. However, when dealing with atoms and subatomic particles, the most convenient unit to use for energy is the electronvolt (symbol eV). One electronvolt is the energy that an electron would gain in moving from one terminal of a one-volt battery to the other terminal. It happens to be just the right sort of size to describe energies at the atomic and subatomic scales. One joule is equivalent to about six billion billion electronvolts.

Now, because mass and energy are interchangeable, it's also convenient to refer to the masses of subatomic particles in terms of their energy equivalence. For instance, the mass of an electron is about 9.1×10^{-31} kg. The energy equivalent of this mass is given by Einstein's equation as around 510 keV (510 kiloelectronvolts or 510,000 electronvolts). The term mass energy is often used to refer to this energy equivalent of the mass of a particle. Similarly, the mass energy of either a proton or neutron is around 940 MeV (940 megaelectronvolts or 940 million electronvolts).

Einstein's equation may also be rearranged as $m = E/c^2$, so masses may be expressed in units of 'energy/c^2'. The approximate masses of an electron and a proton or neutron can therefore be written as 510 keV/c^2 and 940 MeV/c^2 respectively. Whichever units are used, a proton is almost 2000 times more massive than an electron.

Taking the idea of mass-energy equivalence a little further, two ways in which mass and energy can be converted from one to the other are via the processes called pair creation and matter–antimatter annihilation. Although you may have thought antimatter to be the stuff of science fiction, it is a very real feature of the Universe. Antimatter particles have the same mass as their matter counterparts, but their other attributes, such as electric charge, have the opposite sign. All matter particles have corresponding antimatter counterparts, even though the Universe today consists almost exclusively of matter particles rather than antimatter. However, as you will see later, the early Universe was not such a one-sided place.

The antimatter counterpart of the electron, known as the positron (or antielectron), was discovered in 1932. More recently, in 1996, atoms of antihydrogen were created, consisting of antiprotons bound to positrons. Nowadays antimatter particles can be created routinely in high-energy particle accelerators, but antimatter is difficult stuff to control. If matter and antimatter come into contact with each other, they will mutually annihilate, producing a large amount of energy, which appears in the form of electromagnetic radiation. The process of matter–antimatter annihilation may therefore be expressed as:

$$\text{matter} + \text{antimatter} \rightarrow \text{electromagnetic radiation}$$

The process of pair creation is exactly the reverse:

$$\text{electromagnetic radiation} \rightarrow \text{matter} + \text{antimatter}$$

As noted earlier, one of the most important rules governing the Universe is the principle of the conservation of energy: energy cannot be created or destroyed, but merely *changed* from one form to another. So, for instance, any amount of kinetic energy, electrical energy, gravitational energy, mass energy, or energy of electromagnetic radiation may be converted into exactly the same amount of any other type of energy. Notice that this principle implicitly *includes* mass energy as a form of energy, just like any other.

In the annihilation and pair creation reactions, it is important to notice that there are *two* types of energy to consider when talking about the matter and antimatter. First, there is the mass energy of the various matter and antimatter particles, and in addition to this, there is the kinetic energy that the particles possess. So, in annihilation reactions, the energy of the electromagnetic radiation that is produced is equal to the combined mass energy of the matter and antimatter particles *plus* their combined kinetic energy. Similarly, in pair creation reactions, the energy of the electromagnetic radiation can appear as the combined mass energy of the matter and antimatter particles, and any energy left over is imparted to the particles as kinetic energy.

In everyday speech, there is sometimes confusion between seemingly similar concepts such as energy, power, and work. In physics, each of these words has a precise meaning, which are summarized here for completeness. So far, the concept of energy has been discussed in terms of its conservation. Another definition is that the energy of a system is a measure of its capacity for doing work. In turn, the work done on an object by a force is the energy transferred to or from that object by the force. When a force acts on an object, the work done by that force is equal to the change in the object's kinetic energy. Both energy and work are therefore measured using the standard unit of joules. Power is defined as the rate at which work is done and energy is transferred. The standard unit of power is the watt (symbol, W) where 1 watt is equivalent to 1 joule per second.

Finally, it is useful at this point to define what is meant by heat. Again, this word has a familiar usage in everyday speech, but in precise physical terms, heat is a quantity of energy that is transferred between objects as a direct result of a difference in temperature between the objects. Conversely, a quantity of energy transferred between objects *not* as a result of a temperature difference is referred to as work.

2.4   THE WORLD OF LIGHT

Human eyes are sensitive to light that comprises the familiar rainbow of colours – red, orange, yellow, green, blue, and violet – but these colours are merely a tiny part of the vast electromagnetic spectrum. Light and other electromagnetic radiation are the main way that we experience the Universe, and so an appreciation of them is of course necessary to understand the Universe.

When a beam of white light, such as that from the Sun, is passed through a glass prism, it is broken up, or dispersed, into a band of colours that is called the visible spectrum. The spectrum of sunlight contains visible light of all colours and therefore forms a continuous spectrum. Some other objects also emit continuous spectra of light. For instance, the heated metal of a tungsten filament in a light bulb, or the hot-plate on an electric cooker, each emit a continuous spectrum.

An alternative way to disperse light into its constituent colours is to use a device called a diffraction grating. This is simply a flat piece of glass, or other substance, on which are ruled many fine parallel lines – typically a few hundred lines per millimetre are sufficient to disperse visible light. Red light is diffracted through a larger angle than violet light (and all the other colours have a range of diffraction angles in between), so the grating can disperse white light into a spectrum, just like a prism does.

The reason that some things appear to be coloured is that atoms and molecules absorb and emit different colours of light. So, the petal of a rose, for instance, appears red when sunlight shines on it because the petal absorbs all the visible colours *except* the red light. Some of the red light passes through the petal, and some is reflected from its surface.

It turns out that every type of atom shows a preference for absorbing or emitting light of certain *specific* colours. So, each atom is associated with a characteristic pattern of colours – a sort of technicolour fingerprint. Just as every human being has their own set of characteristic fingerprints, so every type of atom has its own pattern of colours of light that it can absorb or emit.

The yellow glow produced by sodium street lights when an electric current passes through vaporized sodium is the fingerprint of that element; similarly, the bright orange–red light of a neon sign is the fingerprint of the element neon. Such associations are even more apparent when a prism is used to disperse the light from a source of known chemical composition. In this case, spectra consisting of emission lines are seen. Exactly the same association may be seen by observing how white light is *absorbed* by atoms. This is done by passing a beam of white light through some vapour and examining the spectrum of the emerging beam to see if any colours are missing. The resulting absorption spectrum exhibits dark absorption lines, marking the absence of exactly the same colours that were seen in the emission spectrum (Figure 2.2). Sodium and neon atoms are by no means unique in having a characteristic pattern of spectral lines. In fact, every kind of atom has an associated, characteristic "spectral fingerprint," and you will see why this is so in Chapter 3.

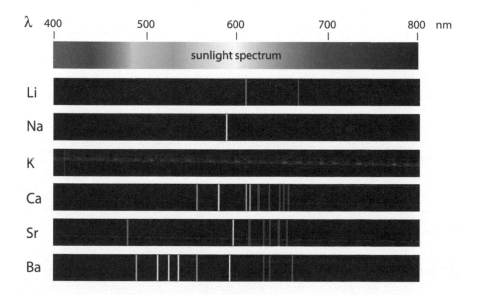

FIGURE 2.2 Emission line spectra from different elements: lithium (Li), sodium (Na), potassium (K), calcium (Ca), strontium (Sr), and barium (Ba).

The emission and absorption of light and other electromagnetic radiation by clouds of gas and dust, by stars, and by galaxies are the key means by which we can understand what the Universe is made of, and also measure its physical properties such as temperature and density. If it were not for the fundamental link between atoms and electromagnetic radiation, we would have very little idea about the composition and nature of the Universe.

One way to quantify the colours in the spectral fingerprint of each element is to use the wave description of light. While it is fairly easy to visualize water waves, or even sound waves, it is not immediately apparent what is "waving" in the case of a light wave. This will be discussed further in Chapter 11, but for now simply note that the distance between one crest of a wave and the next is known as the wavelength of the wave, measured in units of metres or suitable (sub-) multiples of that unit. Wavelength is often represented by the Greek letter λ (lambda). The frequency of a wave is the number of cycles of a wave that pass a fixed point per second, measured in units of hertz (represented by the symbol Hz or s^{-1}). Frequency is often represented by the Greek letter ν (nu). Electromagnetic radiation travels at the speed of light (about 300,000 kilometres per second), and the speed of propagation is simply equal to the frequency of the wave multiplied by its wavelength.

The key thing to remember is that each colour of light corresponds to electromagnetic waves of a different wavelength. Visible light spans the range of wavelengths from about 400-billionths of a metre (400 nanometres, written as 400 nm or 400×10^{-9} m) to 700-billionths of a metre (700 nanometres, written as 700 nm or 700×10^{-9} m). Violet light has the shortest wavelength and red light the longest wavelength. However, at even longer wavelengths are found first infrared radiation, then microwaves, and then radio waves, each of which spans a characteristic range of wavelengths. Infrared radiation stretches from wavelengths of less than a micrometre (1 μm = 10^{-6} m) to about a millimetre (1 mm = 10^{-3} m); microwaves span the wavelength range from about a millimetre to about a metre; and radio waves may have wavelengths from less than a metre to many thousands of metres. At wavelengths shorter than the visible are found ultraviolet radiation, then X-rays, and then gamma-rays. The wavelengths of these types of radiation are so small they are difficult to comprehend. A wavelength of one nanometre (1 nm = 10^{-9} m) lies around the middle of the X-ray range.

Although electromagnetic radiation travels from place to place like a wave, it is emitted or absorbed by matter as if it is composed of a stream of particles, called photons. Monochromatic light, which has a single colour, consists of identical photons, each of which carries exactly the same energy. The energy of a photon depends directly on the frequency used to characterize the electromagnetic wave: higher frequency electromagnetic waves are composed of higher energy photons.

The unit of energy used to quantify a photon's energy is the electronvolt, which was introduced earlier, and conveniently photons of red light each have an energy of just less than 2 eV, while photons of violet light each have an energy of just over 3 eV. At lower energies than red light are the infrared, microwave, and radio wave parts of the electromagnetic spectrum, where photon energies are tiny fractions of an electronvolt (Figure 2.3). At higher energies are the ultraviolet, X-ray, and

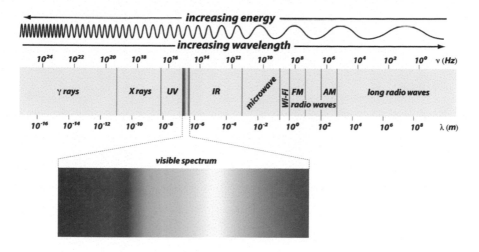

FIGURE 2.3   The electromagnetic spectrum.

gamma-ray regions. X-ray or gamma-ray photons have energies of many kiloelec-tronvolts (1 keV = 10^3 eV), many megaelectronvolts (1 MeV = 10^6 eV), or many gigaelectronvolts (1 GeV = 10^9 eV).

There is no conflict between the two approaches to describing light; it is just that one picture (waves) is useful for describing the way electromagnetic radiation propa-gates and another picture (particles) is useful for describing the way it interacts with matter. This double description is referred to as wave–particle duality.

The energy of a photon is equal to the frequency used to characterize the elec-tromagnetic wave multiplied by a number known as the Planck constant which has a value of about 4.1×10^{-15} eV s. In this sense, the Planck constant can be thought of as a simple conversion factor between photon energy and electromagnetic wave frequency. More fundamentally, the Planck constant lies at the heart of understand-ing the behaviour of matter on its smallest scales. Its introduction in the late 19th and early 20th centuries was a symbol of the revolution in scientific thinking that was necessary for scientists to tackle the problem of understanding the true nature of matter and how it interacts with electromagnetic radiation. The revolution was that of quantum physics, which you will be learning a good deal about in the next few chapters.

Part I

The Small-Scale Universe

As noted in Chapter 1, this part of the book is where the ideas of quantum physics are introduced. In the history of science, new insights have sometimes required great voyages of discovery, such as those of Charles Darwin or Alexander von Humboldt. The discoveries of quantum physics required no such journey as the quantum realm is literally all around us and may be found by travelling only the tiniest fraction of a millimetre into the heart of atoms. Despite this, by undertaking the shortest of all journeys, scientists have uncovered a new quantum world that displays a whole range of features which no one had predicted. This first part of the book therefore addresses the first part of the question posed in Chapter 1 – how does the Universe behave on small scales?

3 Quantized Energy

One of the most revealing ways of investigating atoms is to look at how they interact with light. By examining the light emitted and absorbed by atoms, scientists have been able to draw up a remarkably detailed picture of how the electrons in atoms are arranged. In this chapter you will see that, because atoms are characterized by the light they emit and absorb, it is possible to work out, from the energies of this light, the possible values of energy that the atom can gain or lose.

3.1   THE QUANTUM WORLD

If you had told most physicists at the end of the 19th century that their branch of science was going to undergo a complete upheaval, they would probably not have believed you. At that time, there was a widespread, though not universal, complacency in the scientific community. As early as 1875, a young student called Max Planck began his studies at the University of Munich, where he was supposedly encouraged not to study science as there was nothing new to be discovered! Planck wisely ignored his professor's advice and went on to make what is now regarded as one of the most important discoveries in the entire history of science, a discovery that has been fundamental to the modern understanding of atoms.

Towards the end of the 19th century, several scientists were trying to understand the properties of the radiation emitted by hot objects. *All* objects emit electromagnetic radiation, and the amount of radiation that they produce at different wavelengths depends on the temperature of the object. For instance, the electromagnetic radiation emitted by the Earth is mainly in the form of infrared radiation, whereas that from a hotter body, such as the Sun, is mainly in the visible part of the spectrum. Some people thought this phenomenon was a minor diversion that would be understood using existing ideas. However, it turned out to be a far more subtle process than most people assumed and led to nothing less than a scientific revolution.

In 1900, Planck made the astonishing suggestion that when objects emit or absorb radiation, they can do so only in multiples of a certain minimum amount of energy called a quantum (plural quanta, from the Latin word *quantus* meaning "how much"). So, for example, the gas inside an energy-saving light bulb emits energy in the form of light, but it does not emit energy in just any amounts, rather it emits it in certain, particular amounts: quanta of light. Likewise, when light is absorbed by, say, a piece of metal, the energy of the light is absorbed only in quanta. The mathematical details of Planck's idea are not relevant, but what is important for the present purposes is the historical significance of his idea. Planck was the first to introduce the idea of the quantum into science, but within 25 years, other scientists had used it to form the basis of quantum physics, which in turn is the basis of the current understanding of how atoms behave.

This scene-setting for your exploration of quantum physics concludes with a consideration of the vital importance of this subject for understanding and describing nature. If you ask an astronomer why stars, such as the Sun, shine so brightly, you will learn that the answer lies ultimately in the realm of quantum physics. If you ask a chemist why some atoms stick together to form molecules but others do not, or ask a biologist why DNA molecules fold up in the particular way that they do, you will again be told that the answer ultimately lies in quantum physics. If you ask almost anyone how the whole Universe came into existence they will probably say that they haven't a clue; but if you speak to one of the scientists who are actively engaged on this problem they will tell you that their best bet is that it involves quantum physics in some fundamental way.

Quantum physics explains how the fundamental particles of matter and radiation interact with one another. It explains how and why electrons are arranged the way they are in atoms, and this in turn provides an explanation for the various properties of the different elements. Ultimately, therefore, all the rich diversity of chemical reactions relies on the rules of quantum physics. Taken further, the chemical reactions that underlie biological systems, such as the transcription of DNA and the process of glucose oxidation, also rely on the underpinning rules of quantum physics. Some scientists even suggest that the essence of consciousness may lie in quantum processes occurring within human brains.

At the end of Chapter 4, the theme of the relevance of quantum physics to everyday life will be revisited and a quick overview will be presented of how quantum physics has enabled the technology that many people now take for granted.

3.2   THE SPECTRAL FINGERPRINT OF HYDROGEN

Photons can reveal a great deal about the inner workings of an atom. Each spectral line involves the emission or absorption of photons of a particular energy. So it follows that in order to explain why each type of atom is associated with a particular set of spectral lines, what is really needed is an explanation of why each type of atom only emits or absorbs photons that have certain fixed amounts of energy. These photon energies are just as characteristic of a given type of atom as the pattern of spectral lines to which they correspond. In fact, each type of atom can be characterized by the energies of the photons that it can absorb or emit.

The message is simple: every atom has a spectral fingerprint that can be specified in terms of either the colours of the radiation it emits or absorbs or, more precisely, the energies of the corresponding photons. You will now see how to apply these ideas to the simplest atom of all, the hydrogen atom.

The hydrogen emission spectrum contains five visible lines at energies of 1.89 eV (red), 2.55 eV (blue-green), 2.86 eV (deep blue), 3.02 eV (violet), and 3.12 eV (also violet) (Figure 3.1). This set of energy values represents the experimental fact that hydrogen atoms emit visible light only with these five values of energy (likewise, they can absorb visible light only with these five energy values too). So, for example, a hydrogen atom can emit a photon with an energy of exactly 1.89 eV, but it can never emit a photon with an energy of, say, 1.91 eV. The hydrogen atom is very particular

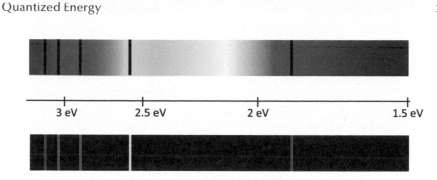

FIGURE 3.1   The emission line spectrum and absorption line spectrum of hydrogen.

about the light it emits and absorbs. The same thing is true for all other atoms – every type of atom can emit or absorb photons with only particular values of energy.

3.3   THE ENERGY OF AN ATOM

You know that energy is always conserved. This implies that if an atom *absorbs* a photon that has a given energy, then the energy of that atom must *increase* by the same amount of energy. Similarly, if an atom *emits* a photon of a particular energy, then the energy of that atom must *decrease* by the same amount of energy. So, the fact that a given type of atom can absorb and emit photons of certain, precisely defined, energies must mean that the atoms themselves are only able to increase and decrease their own energy by exactly those amounts.

We now consider what is meant by "the energy of an atom." First, you should not confuse this with the kinetic energy of the atom as a whole. The atom may well be moving, and so possess a certain amount of kinetic energy, but changes in this energy are not responsible for atomic spectra. Rather the concern here is with what may be thought of as the *internal energy* of the atom. Now, the internal energy of a liquid, say, has two components: the *kinetic energy* of the random motion of the constituent molecules and their *potential energy* that results from the molecular forces of attraction between the molecules. A similar situation applies in the case of individual atoms. The simplest case, a hydrogen atom, consists of a single electron bound to a single proton. The electron will have a certain amount of kinetic energy (because it is moving), and there will be a certain amount of electrical potential energy due to the electrical force of attraction between the negatively charged electron and the positively charged proton. The *total* (internal) energy of the atom is the sum of these two contributions.

Now think about what will happen if the electron in a hydrogen atom is moved further away from the proton. For comparison, as an object is raised above the surface of the Earth, work is done against the gravitational force of attraction between the object and the Earth. As a consequence of this, the object acquires increased gravitational potential energy. Here, as the electron is moved further from the proton, work is done against the electrical force of attraction between the two. So, in this case, there is an increase in *electrical* potential energy. At the same time, when an electron is moved further from the proton it will generally move more slowly, so its *kinetic* energy will decrease. In fact, the electrical energy generally has a larger magnitude than the kinetic energy, so the electrical energy dominates the energy of the atom. In general, therefore, the further away the electron is from the proton in a hydrogen atom, the greater will be the energy of the atom.

So when a hydrogen atom absorbs or emits photons, the energy of the atom will alter and changes will occur to the internal structure of the atom. In particular, the position and velocity of the electron will change. When a photon is *absorbed*, the position and velocity of the electron will change so that the atom has a *higher* energy, and when a photon is *emitted*, the position and velocity of the electron will change so that the atom has a *lower* energy.

As you will see in Chapter 4, the whole question of the position and velocity of an electron in an atom is subject to so-called quantum indeterminacy, so the picture outlined above is only a guide to the true situation. Nonetheless it is a useful one to bear in mind, and similar conclusions apply to more complex atoms.

3.4   ENERGY LEVELS AND TRANSITIONS

The ability of any given type of atom to absorb or emit only certain characteristic amounts of energy is explained by saying that the atom itself can have only certain values of energy, known as energy levels. The energy is said to be *quantized*, because the energy levels correspond to only certain, well-defined, values of energy. An atom with a particular value of energy, corresponding to a certain energy level, can change to another energy level by absorbing or emitting a photon (there are other ways too, for example by colliding with other atoms), and because energy is conserved, the energy of the photon is precisely equal to the energy difference between the two energy levels. When the atom makes such a change it is said to undergo a transition to a different energy level. Such transitions are often informally referred to as quantum jumps. This is the first of two key features of the quantum world of atoms: each type of atom has a different set of quantized energy levels. When atoms emit or absorb photons, they make transitions between these quantized energy levels.

The emission and absorption spectra that you saw earlier arise simply as a result of these transitions between energy levels. When a large number of sodium atoms, for instance, make transitions from a certain high energy level to a certain *lower* energy level, photons of a particular energy are *emitted*. When white light is directed at a large number of sodium atoms, some of them will make transitions from a certain low energy level to a certain *higher* energy level, and photons of a particular energy are *absorbed*.

Now, the numerical value of the energy of an atom that corresponds to each energy level is not generally known. Rather it is the *differences* in energies between *pairs* of levels that are well defined, because it is these differences that correspond to the energies of the photons that the atom may absorb or emit. This is just the familiar principle of conservation of energy that you have met before, and can be summarized by saying that when an atom absorbs or emits a photon, the photon energy is equal to the value of the higher energy level minus the value of the lower energy level.

In the case of sodium atoms, the two energy levels involved in the emission or absorption of the yellow line in the spectrum (shown in the previous chapter) are separated by an energy difference of 2.1 eV. The energy of the photons of yellow light in the sodium spectrum is therefore 2.1 eV.

3.5   THE ENERGY LEVELS OF THE HYDROGEN ATOM

You have seen that hydrogen atoms emit and absorb visible light of just five specific energies: 1.89 eV, 2.55 eV, 2.86 eV, 3.02 eV, and 3.12 eV. It is important to remember that these are only the visible lines – hydrogen emits and absorbs radiation that is not visible, for example in the infrared and ultraviolet regions of the electromagnetic spectrum, too. In order to build up a complete picture of the energy levels of the hydrogen atom, you would need a complete list of all the energies of the photons that it can emit or absorb.

Scientists have indeed produced such a list, and this allows the complete energy-level diagram for the hydrogen atom to be drawn, in which the energy values are illustrated in a conventional and convenient way (Figure 3.2). What matters here is the vertical spacing of the energy values; neither the horizontal extent of the lines nor their thickness is of any significance whatsoever. One important thing to notice about this energy-level diagram is its simplicity: the pattern is one in which the energy levels gradually come closer together. The higher energy levels of hydrogen become so densely packed that it becomes impossible to draw them as distinct lines and the best that an illustrator can do is to indicate that the lines gradually get closer and closer together. The energy-level diagram for hydrogen is in fact the simplest of all atomic energy-level diagrams, just as you might have supposed.

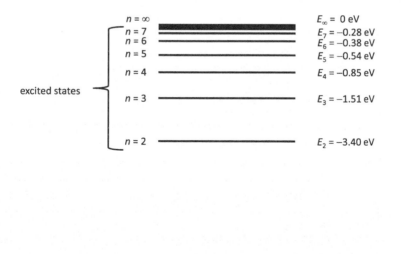

FIGURE 3.2   The hydrogen energy-level diagram.

You can imagine the energy levels in a hydrogen atom as being a little like the rungs of a ladder that is sunk into a deep pit. In this analogy, the lowest rung of the ladder (the lowest energy level) is near the bottom of the pit, just above where the nucleus of the atom sits. When the hydrogen atom occupies its lowest energy level, you can imagine the electron sitting on this lowest rung, close to the nucleus. As the hydrogen atom occupies higher and higher energy levels, the electron sits on ever higher rungs of the ladder, until eventually the atom has so much energy that the electron climbs beyond the top rung and moves away from the pit.

In an energy-level diagram, the relative separations are usually drawn to scale, so that each of the photon energies associated with hydrogen is represented by the distance *between* two energy levels. For instance, photons of visible radiation (often called simply visible photons) with the *lowest* energy (1.89 eV) are associated with transitions between the second and third energy levels, whereas the visible photons with the *highest* energy (3.12 eV) are associated with transitions between the second and seventh energy levels. More widely separated levels indicate transitions of relatively large energy, and so photons associated with these transitions may correspond to the ultraviolet part of the spectrum (energy differences greater than about 3.2 eV). Conversely, closely separated levels indicate transitions of relatively small energy, and so photons associated with these transitions may correspond to the infrared part of the spectrum (energy differences less than about 1.5 eV). Remember, transitions do not just occur between adjacent energy levels, but between *any* pair of energy levels.

When a hydrogen atom (or in fact any other atom) sits in the *lowest* energy level, it is said to be in its ground state. This is the "normal" state in which an atom might be found. It will correspond to a situation in which the electron is likely to be close to the proton, with a minimum value of electrical energy. Above this energy, when the atom sits in its second, third, fourth, etc., level, it is said to be in an excited state. In principle there is an infinite number of energy levels, and the topmost one corresponds to a situation in which the electron and proton are widely separated, with a maximum value of electrical energy.

The next point to note is that although an energy-level diagram may contain an infinite number of levels there is still only a finite difference in energy between the lowest energy level and the highest energy level. This shows that there is a limit to the amount of energy that a hydrogen atom can absorb without the electron and proton being split apart. If the atom is initially in its ground state and it absorbs a photon with an energy greater than the difference between the ground state and the uppermost energy level, the electron is freed completely from its bondage to the proton at the core of the atom (the nucleus). The removal of an electron from a hydrogen atom leaves a charged particle which is written H^+ and is called an ion. (In fact, the hydrogen ion is simply a proton.) The technical term for such removal of an electron from an atom is therefore ionization. The energy difference between the topmost energy level and the ground state is called the ionization energy, which in the case of hydrogen is about 13.60 eV.

There is a continuous range of possible energies above the ionization energy, and this is generally referred to as the continuum. The continuum corresponds to situations in which the electron and proton move around separately, with a combined kinetic energy that is greater than the topmost energy level. The electron is no longer bound to the proton, and the particles can have *any* value of energy.

Now for a rather subtle point. You have already seen that the energies of the levels are not really known. However, there is no such doubt about the *differences* in energy between the levels. These differences are very well defined, and, as you have already seen, they can be determined quite simply from atomic spectra. The energy of any of the energy levels is not known, but it makes life simpler for everyone if it is agreed, as a matter of convention, that the energy of the topmost energy level is 0 eV. This corresponds to the energy of a separated electron and proton. All bound states of atoms, therefore, correspond to *negative* values of energy levels, as these energy levels sit below the topmost energy level. So, the ground-state energy is 13.60 eV below that, namely at −13.60 eV. Furthermore, each of the excited levels can then be associated with some negative value of energy between −13.60 eV and 0 eV. This is the convention that is adopted for the rest of this book. Note that this is not implying that the atom has a "negative energy" when it is in one of the bound states shown, simply that it has *less* energy than if the electron and proton were separated.

There is one final point to make about the energy-level diagram of hydrogen. If you carefully examine the energies of the levels, you will find that the energy associated with the nth energy level is given by -13.60 eV$/n^2$ where $n = 1, 2, 3 \ldots$ etc. This remarkably simple formula provides a wonderfully compact way of remembering everything that is shown in the energy-level diagram of hydrogen. The formula applies only to the hydrogen atom; indeed, it is not possible to summarize all the energy levels of any other atom in a simple formula.

Given the full range of hydrogen energy levels, it is possible to identify *all* possible transitions between energy levels in the hydrogen atom and therefore all the possible spectral lines. In the case of the hydrogen atom these transitions are grouped into series, named after the scientists who discovered them. For instance, transitions to or from the first energy level give rise to lines of the Lyman series in the ultraviolet; transitions to or from the second energy level give rise to lines of the Balmer series in the visible (including the five lines mentioned earlier); and transitions to or from the third energy level give rise to lines of the Paschen series in the infrared.

3.6   QUANTUM STATES AND QUANTUM NUMBERS

As noted above, when the hydrogen atom is in its lowest energy level then it is in its ground state, and when it is in a higher energy level then it is in an excited state. A general name for these states of the atom is quantum states (or just *states* for short), and each quantum state has a precisely defined amount of energy – one of the energy levels of the atom. However, it turns out that there is *more than one* quantum state corresponding to each value of n in the hydrogen atom, and therefore associated with the nth energy level. The energy is virtually the same for all quantum states that correspond to a given value of n, but the possible values for the position and velocity of the electron in one state will be different from the possible values for the position and velocity of the electron in another state with the same value of n.

Before exploring these quantum states and how they may be described, here is one final piece of terminology: in the hydrogen atom, the *electron* can be said to *occupy* a particular quantum state of the atom. The reason for this is that it is only the possible values of the position and velocity of the *electron* that vary from one quantum state to another – the nucleus of the atom is unchanged. So, when a hydrogen atom undergoes a transition from one quantum state to another, the position and velocity of the electron after the transition will be different from those before the transition. The crucial point is that energy levels and quantum states apply to atoms as a whole, but in the hydrogen atom, the (single) electron can be said to *occupy* a particular quantum state.

It turns out that there are 2 quantum states associated with the first energy level, 8 quantum states associated with the second energy level, 18 quantum states associated with the third energy level, and 32 quantum states associated with the fourth energy level. There is a pattern here: there are $2n^2$ different quantum states associated with the nth energy level of hydrogen.

To reiterate, each of the $2n^2$ quantum states associated with a given energy level will have *almost* the same energy, and for the purposes of this book it is generally sufficient to assume that they all do indeed have the same energy. However, the possible values for the position and velocity of the electron when it occupies each of those quantum states will be different from the possible values in another quantum state of the same energy.

The fact that there are different quantum states associated with a single energy level means that, to label a given state of a hydrogen atom, more is needed than just the number n that labelled the energy levels. Consequently, each of these different quantum states can be characterized by a unique set of quantum numbers. The number n is called the principal quantum number of the state, and it can be any positive whole number (integer): 1 or 2 or 3 or 4, etc. To a good level of accuracy, the principal quantum number n determines the energy of any given quantum state in the hydrogen atom, according to the rule stated earlier. All states with a particular value of n have virtually the *same* energy in the hydrogen atom, given by the formula: -13.60 eV/n^2

To specify a state completely, however, the values of three more quantum numbers are generally needed. The first of these new quantum numbers is called the orbital quantum number and is usually given the symbol l. When used to distinguish states that share a common value of n, it is only allowed to take the whole number values from 0 up to $n - 1$. For example, any state with $n = 1$ (the lowest allowed value of n) must have $l = 0$ (because $n - 1 = 0$), but among the eight different states with $n = 2$ there will be some with $l = 0$ and others with $l = 1$ (the maximum value of l because $n - 1 = 1$). Among the 18 states with $n = 3$ there will be some with $l = 0$, some with $l = 1$ and some with $l = 2$.

The two quantum numbers n and l still don't uniquely identify all the possible states corresponding to a particular energy level. The third quantum number is known as the magnetic quantum number, and it can have any whole number value between $-l$ and $+l$. It is represented by the symbol m_l. So, when $l = 0$, $m_l = 0$ only, but when $l = 1$, $m_l = -1$, 0, or $+1$ and when $l = 2$, $m_l = -2$, -1, 0, $+1$, or $+2$, for instance.

This gives a total of $(2l + 1)$ different values of m_l for any given value of l. Finally, the fourth quantum number m_s is known as the spin quantum number, and it can be either $+1/2$ or $-1/2$ only.

The existence of $2n^2$ different quantum states corresponding to each value of n (and therefore to each energy level) can now be fully accounted for in terms of the rules that have been introduced, namely:

1. The principal quantum number, n, may be any positive whole number (1, 2, 3, etc.), and this number determines the energy level of the hydrogen atom.
2. For a given value of n, the orbital quantum number l may take any whole number value from 0 up to $n - 1$.
3. For given values of n and l, there are $(2l + 1)$ different values of the magnetic quantum number, m_l, ranging from $-l$ to $+l$.
4. For each combination of n, l, and m_l there are two possible values of the spin quantum number, m_s.

As an example, consider the quantum states with an energy that corresponds to the first excited energy level (i.e. the second energy level) of the hydrogen atom. All of these states will have $n = 2$ (Rule 1), so the only values of l they are allowed to have are 0 or 1 (Rule 2). How many of these states will there be in total? Well, using Rule 3, m_l will be 0 when $l = 0$ and it will be -1, 0, or $+1$ when $l = 1$. Finally, from Rule 4, in each case m_s can be $+1/2$ or $-1/2$. So, there are 2 states with $l = 0$, and 6 states with $l = 1$. The total number of possible states is therefore $2 + 6 = 8$. This is exactly the number predicted by the $2n^2$ rule you met earlier.

It is rather tedious to keep saying "a state with $n = 2$ and $l = 0$," to distinguish it from "a state with $n = 2$ and $l = 1$," so a different convention is commonly used among scientists. This involves identifying the value of l by means of a letter, with "s" for 0, "p" for 1, "d" for 2, and "f" for 3. These letters once stood for "sharp," "principal," "diffuse," and "faint" and were coined by scientists examining the spectral lines produced by transitions between these states. The convention is to show the values of n and l that correspond to a particular state by writing the numerical value of n and the letter representing l next to each other. Thus, the ground state of hydrogen, with $n = 1$ and $l = 0$, is a 1s state; an excited state of hydrogen with $n = 2$ and $l = 1$ simply becomes a 2p state; and a 4s state would be an excited state in which $n = 4$ and $l = 0$. This so-called spectroscopic notation is therefore a useful shorthand.

Before leaving the energy-level diagram and quantum states of hydrogen, consider one last feature of this most ubiquitous of atoms. A hydrogen atom in its ground state will sit in one of the two available 1s quantum states. The two possible sets of quantum numbers (n, l, m_l, m_s) that the atom can have are $(1, 0, 0, +1/2)$ and $(1, 0, 0, -1/2)$, and the energy difference between these states is a tiny six-millionths of an electronvolt. When a hydrogen atom makes a spin-flip transition between these two quantum states, it emits (or absorbs) a radio wave with a wavelength of 21 cm. The 21-cm radio emission has proved to be a vital tool in mapping the Milky Way. Most of the Universe is composed of hydrogen, and the structure of the Milky Way is delineated by the clouds of hydrogen gas that it contains. By tuning radio telescopes to pick up the 21-cm wavelength, astronomers can trace out where the hydrogen gas lies, and so unravel the structure of the local patch of space. It is worth remembering that this is only possible because the 1s energy level of the hydrogen atom is split into two distinct quantum states, separated by a few millionths of an electronvolt in energy.

3.7   ENERGY LEVELS IN GENERAL

The focus so far has been on the energies of atoms and you have seen that all atoms have energy levels. This is the remarkable feature that is new and different about quantum physics. In the "everyday physics" you may be familiar with, there is nothing to prevent an object from having *any* amount of kinetic energy, or gravitational potential energy, or electrical energy, or any other kind of energy. The energy of everyday objects, such as cars, people, or grasshoppers, can vary continuously and take any value you care to choose. However, down at the level of the quantum world of atoms, things are very different.

It is now time to look beyond atoms, to apply these ideas to other aspects of the quantum world. Quantum physics, in fact, makes a bold and very general prediction: whenever particles are bound together, they form something that has energy levels. Atoms are a familiar example of this – because every atom consists of electrons bound together with a nucleus, the atom will have energy levels.

Experiments confirm these predictions of quantum physics: every day in laboratories all over the world, scientists study the spectra of molecules and of nuclei. Molecular spectral lines (typically in the infrared and microwave parts of the spectrum) correspond to transitions between different vibrational or rotational quantum states of the molecule, which are quantized. Nuclear spectral lines (typically in the gamma-ray part of the spectrum) correspond to transitions between different quantized states of nuclei. The idea of energy levels, so completely foreign to experiences in the everyday world, is a routine part of the quantum world. In summary: every system in which particles are bound together, such as nuclei, atoms, or molecules, will have *quantized* energy levels. Such systems are referred to as quantum systems.

After this discussion of the line spectra of atoms, you may be wondering how it's possible to get a continuous spectrum, such as that shown earlier. Continuous spectra contain a continuous distribution of photon energies. They can be produced, for example, by switching on an old-style tungsten filament light bulb (which heats the tungsten filament to a very high temperature), or by heating a plate on an electric cooker to a somewhat lower temperature. The red glow from a hot-plate is not attributable to any particular photon energy; this spectrum is continuous, like that of the light bulb, except that the brightest part of the hot-plate's visible spectrum is the red part. Both these devices also emit radiation in other parts of the electromagnetic spectrum, such as infrared radiation.

The continuity of the spectrum from a heated object results from the fact that this is not emission from individual atoms, but the effect of many atoms together in a solid. In a solid metal, like tungsten, the atoms are arranged in a regular fashion, and some of the electrons are shared by the whole array of atoms. This is what makes the conduction of electricity possible. Though highly mobile, these electrons are confined, or bound, within the metal, so they are associated with energy levels. However, there are so many levels, and their energies are so close together, that they form an apparently continuous energy band that is typically a few electronvolts wide. Transitions within this band give rise to a continuous range of photon energies, and so produce a continuous spectrum. The energy levels of a metal, therefore, provide yet another example of quantized energy, in addition to the molecular, atomic, and nuclear energy levels discussed above.

4 Quantum Uncertainty

One of the great attractions of Newton's laws of motion, mentioned earlier, is that they deal in certainties. By using his laws of motion and gravity, everything in the Universe from the period of a pendulum's swing to the motion of the planets can be predicted with a complete and reassuring certainty. Yet scientists now know that it is simply not true that if you know the present state of something, you can necessarily predict its future with certainty. Never mind the entire Universe, it is not even possible to predict every aspect of the future of a single hydrogen atom!

You saw in the previous chapter how Planck's ideas shed light on atomic energies by showing that these energies have only certain, allowed values known as energy levels. But *why* do atoms have energy levels? This question could not be answered using the laws of physics that were known at the beginning of the 20th century. It became clear to a few young and exceptionally gifted scientists that a new approach was needed to describe the behaviour of atoms.

So, in the mid-1920s, quantum physics was born. In quantum physics it is necessary to come to terms with intrinsic "uncertainties" or, more specifically, *indeterminacies*. As you will see in this chapter, quantum physics says there are some things about atoms that are indeterminate, that is, they cannot be known or determined, no matter how clever you are or how much computer power you have to hand. For instance, it is impossible to say exactly *where* an electron in an atom is at a certain time and simultaneously how *fast* it is moving or in which *direction*: the position and velocity are indeterminate.

Most people find it difficult to come to terms with the idea of quantum indeterminacy: Albert Einstein was the most famous of those who objected to the idea. He believed that it must be possible, if you have a good enough theory and good enough equipment, to probe an atom in as much detail as you like. But, despite his brilliant arguments, nearly all quantum scientists believe that on this issue he was wrong. So be prepared to spend time getting to grips with indeterminacy, one of the most challenging concepts in quantum physics.

4.1 INDETERMINACY AND PROBABILITY

When dealing with transitions between energy levels that require an increase in the energy of an atom, it is possible to exert some degree of control over which transitions occur by regulating the energies of the photons that are supplied to the atom. For example, suppose that a hydrogen atom is initially in its ground state. You can ensure that the only upward transition this atom makes is to the third energy level by only allowing the atom to interact with photons whose energy is precisely equal to the energy difference between the two levels. These photons carry just the right amount of energy to cause the desired transition, but too little to cause the jump to the fourth energy level, and too much to cause the jump to the second energy level. Even so, quantum physics allows the atom a degree of unpredictability in that although the atom *may* absorb a photon that has just the right energy, it is not *required* to do so. In a large number of identical encounters between hydrogen atoms and photons with this energy, all that can be predicted is the probability that the photons will be absorbed.

The significance of probability in quantum physics is even more apparent if you consider downward, rather than upward, transitions. Once again consider a hydrogen atom, this time initially in the third energy level. Such an atom may spontaneously make a downward transition, giving out a photon in the process. If it does so, that transition may involve the relatively small jump down to the second energy level and the emission of a 1.89 eV photon of red light, or it might involve the much bigger jump down to the first energy level and the emission of a 12.09 eV photon of ultra-violet radiation.

Which transition will any particular atom make? The answer is that no one knows and, more importantly, according to quantum physics, *no one can know*. Given a large number of identical hydrogen atoms all in the third energy level, it's possible to predict the proportion of them that will emit ultraviolet photons and jump to the first energy level, and the corresponding proportion that will emit visible photons and jump to the second energy level, but it is *impossible* to predict which of the two possible jumps will be made by any particular atom. This idea can be summarized in the following way: in quantum physics, the possible outcomes of a measurement can generally be predicted, and so can the probabilities of each of those possible outcomes. However, it is not generally possible to definitely predict the outcome of any individual measurement if there is more than one possible outcome.

A rather extreme illustration of quantum indeterminacy was raised in an article published by Erwin Schrödinger in 1935. He imagined that a cat is placed in a closed box containing a device which consists of a single radioactive nucleus, a radiation detector, and a phial of poison gas (Figure 4.1). These are connected in such a way that if the nucleus decays, the detector registers the emitted radiation, the phial will be smashed, and the gas will be released, so killing the cat. Initially the nucleus has not decayed and the cat is alive. However, due to the indeterminacy of quantum physics, once the box is closed it is impossible to say whether the nucleus has decayed or not, until it is observed to have done so. In one sense it is in *both* states simultaneously. The nucleus is said to be in a superposition of quantum states – in one of which it has decayed and in the other of which it has not. Likewise, the cat too will be in a superposition of quantum states – in one of which it is alive and in the other of which it is dead. Schrödinger's cat is somehow both alive and dead at the same time.

The implications of quantum physics, such as those exemplified by the poor cat just discussed, are sometimes difficult to comprehend. Nonetheless, it is believed that in quantum physics even the most completely detailed description of a system, for example an atom in a particular energy level, will still only allow predictions to be made about the probabilities of different outcomes for the future behaviour of the system. In other words, in quantum physics, the use of probability is an *essential* feature and not simply a matter of convenience or practicality. Quantum physics,

FIGURE 4.1   Schrödinger's cat.

with the built-in imprecision of a description of nature based on probabilities, is supported by experiment. Not a single experiment has ever disproved quantum physics.

In quantum physics therefore, there is a distinctive combination of strong prohibition and unpredictable freedom: whatever is not actually forbidden will be found to occur, sometime or other. In other words, what can happen will happen, sooner or later! The fact that even the most detailed possible description of a quantum system still involves probability in an essential way shows that indeterminacy is built into the heart of quantum physics.

4.2   MODELLING ATOMS

The structure of an atom cannot be seen directly, so instead scientists find it useful to come up with ways of describing atoms in terms of pictures or mathematical equations. Such a description is usually referred to as a model. If you have any preconceived picture of what an atom looks like, it may be that you imagine the electrons orbiting the nucleus, much as the planets in the Solar System orbit the Sun. Indeed, a model like this was proposed by the Danish physicist Niels Bohr in 1913. His model included fixed orbits for the electrons, each of which corresponds to one of the energy levels. Higher energy levels corresponded to orbits that were further from the nucleus. While this is a simple and appealing picture that can explain some phenomena, it has shortcomings that make it inconsistent with the real world. For instance, if electrons really did orbit the nucleus, then they would be constantly accelerating. Any charged object undergoing an acceleration will continuously emit electromagnetic radiation and lose energy. As a result, an electron orbiting in this way would rapidly spiral into the nucleus as its electrical energy reduced. This does not happen, so electrons cannot really be orbiting the nuclei of atoms.

The richly detailed understanding of the chemistry and physics of atoms that modern science provides is rooted in a quite different model of the atom that fully incorporates the essential indeterminacy of quantum physics. The quantum model of the atom that is used today emerged from the work of a number of European physicists in the mid-1920s. One of the most influential members of this group was the Austrian theoretical physicist Erwin Schrödinger mentioned earlier, so it is convenient to refer to the model as the Schrödinger model of the atom. The model is mathematically complex, but here the focus is only on its basic concepts and results rather than the methods used to obtain them. As usual the discussion concentrates on the hydrogen atom, but it is worth noting that the model can successfully describe all other atoms too. When the Schrödinger model is applied to the hydrogen atom, it predicts that the atom has energy levels given by the formula: -13.60 eV/n^2, presented earlier. It also predicts the existence of each of the $2n^2$ quantum states associated with each energy level, and the allowed values of quantum numbers in each case. Finally, Schrödinger's model predicts detailed information about the possible position and velocity of the electron when the atom sits in each of these quantum states.

The first of these points confirms that the Schrödinger model accounts for the emission and absorption spectra (the "spectral fingerprint") of hydrogen. The final point predicts that the behaviour of the electron in the atom is intrinsically indeterminate. According to quantum physics, if a hydrogen atom sits in a particular quantum state (with a certain value of energy), it is impossible to say in advance of a measurement what values will be obtained for the electron's position or velocity. This means that if *identical measurements* are made on atoms in the *same* quantum state, the experiments will give a *variety of different outcomes*. What *can* be predicted are all the possible outcomes of a measurement (i.e. all the possible positions and velocities of the electron) and the probabilities of each of those outcomes.

It turns out that the most probable position for an electron in a hydrogen atom is further from the nucleus for the higher energy levels, and closer to the nucleus for the lower energy levels. However, a large range of positions for the electron are possible in each energy level. To draw an image of the atom, the best that can be done is to show a "cloud" surrounding the nucleus representing all the possible positions that the electron may have (Figure 4.2). Different quantum states correspond to different sets of possible values for the electron's position, and so will be described by different electron clouds. Some of these clouds are spherical, whereas for some quantum states, the electron clouds are shaped like dumb-bells, clover leaves, or ring doughnuts!

Remember, the electron cloud is *not* a depiction of the positions of many electrons, it's the possible positions of a *single* electron in the atom. In some sense, the single electron in a given quantum state is present in *all* the locations of the cloud simultaneously, including those locations that are separated spatially by regions in which the electron is *never* likely to be found! How the electron gets from one location to

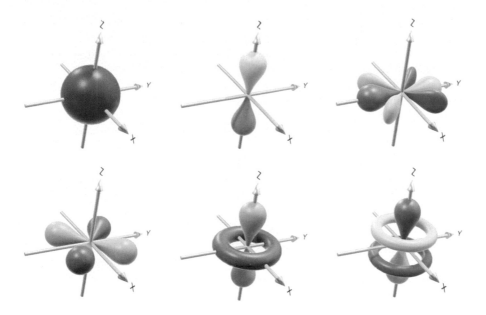

FIGURE 4.2   Electron probability clouds for various hydrogen quantum states.

another without passing through the intervening space is not addressed by quantum physics. Here then is a further demonstration of indeterminacy at work in quantum physics and another illustration of the essential use of probability in understanding how atoms behave.

4.3   THE UNCERTAINTY PRINCIPLE

Although the precise position and velocity of an electron in an atom cannot be *predicted*, in theory it is possible to devise experiments to *measure* its position or velocity. But can *both* the position and velocity be measured simultaneously, so tying down the behaviour of the electron precisely? It turns out that the answer is no. A fundamental result of quantum physics, discovered in 1927 by Werner Heisenberg, explicitly rules out such detailed knowledge of both the position *and* velocity of a particle. The Heisenberg uncertainty principle limits the precision with which one quantity (such as position) may be known when combined with the precision with which certain other quantities (such as momentum, which depends on the particle's mass and velocity) may also be known. This limit is expressed by saying that the uncertainty in the position of a particle multiplied by the uncertainty in its momentum must be greater than the value of the Planck constant divided by 4π.

According to the uncertainty principle, measuring one quantity with a prescribed level of precision automatically limits the precision with which the other can be known at the time of the measurement. Indeed, the uncertainty principle says that if you know the exact position of a particle (i.e. with zero uncertainty), you can know nothing at all about its momentum (and so its velocity) at the same time. The converse is also true: if you know its momentum (or velocity) exactly (i.e. with zero uncertainty), you can't know anything about its position simultaneously.

To get some idea of why this is the case, think about what measuring the position of an electron might actually involve. The obvious way to determine the position is to "see" where the electron is, and to "see" where something is you have to shine light on it. But shining light on an atom will have one of two effects. On the one hand, a photon may be absorbed, so causing the atom to jump to another energy level, in which case the quantum state of the atom has changed. Alternatively, the photons may not interact with the atom, but emerge unaffected, in which case you haven't measured anything! In other words, by measuring the position of the electron in an atom, you will change the energy of the atom, and so alter the possible range of velocities of the electron. Indeed, *any* act of measurement on a hydrogen atom, or any other quantum system, will involve transferring energy into or out of that system, and so will change its properties.

It's important to appreciate that the limitations of the uncertainty principle are a matter of deep principle in quantum physics, not a result of sloppy work or poor equipment. The quantum world is not only essentially indeterminate, it is also inherently uncertain.

Perhaps the ultimate expression of quantum indeterminacy is the fact that the smallest constituents of matter, such as electrons, really do not behave like particles, localized at a point in space, at all. The French physicist Louis de Broglie proposed that, just as electromagnetic radiation propagates like a wave and interacts like a particle, so *any* moving "particle" has an associated wave and will therefore exhibit wave-like behaviour such as diffraction by a suitable grating. The wavelength associated with any particle is equal to the Planck constant divided by its momentum (which in turn is equal to its mass multiplied by its velocity) – so, the smaller the mass and the slower it moves, the larger is its de Broglie wavelength. For instance, the de Broglie wavelength of an electron moving at 1% of the speed of light is about 0.2 nm, and a beam of electrons will therefore be *diffracted* by the regular lattice formed by atoms in a crystal, with a spacing of about 0.5 nm. De Broglie's idea of wave–particle duality may be summed up by noting that electrons interact like particles but propagate like waves.

It is a remarkable irony that the English physicist J.J. Thomson discovered the electron, displaying properties of a particle, in 1897, and that his son, George Thomson, demonstrated electron diffraction, proving that electrons can behave like waves in 1927. Both models of the electron's behaviour are appropriate in different circumstances.

In the everyday world, however, quantum indeterminacy, the Heisenberg uncertainty principle, and de Broglie's wave–particle duality can largely be ignored. The reason for this is that atoms, and other quantum systems, are so very small and have such a tiny mass. The more massive an object, the less important are the effects of indeterminacy. For instance, the uncertainties in the position or velocity of a planet orbiting the Sun are so tiny that they are utterly insignificant, and, in effect, its position and velocity can be measured as precisely as measuring instruments allow.

Likewise, the de Broglie wavelength of a person is so tiny that you are not noticeably diffracted when you walk through a doorway! In general, quantum indeterminacy is only apparent when dealing with things on an atomic scale, or smaller. But it is this very reason why an understanding of quantum effects is necessary to construct the integrated circuits of semiconductor devices that lie at the heart of the technological world. Without knowledge of quantum indeterminacy, the uncertainty principle, and wave–particle duality, the world would be a very different place.

4.4   QUANTUM TECHNOLOGY

This chapter concludes by again considering the fundamental importance of quantum physics, this time by thinking about the technology that is enabled by an understanding of it. Since the 1920s, knowledge about quantum physics has continued to grow and diversify. It provides insights into a vast range of phenomena and supplies the scientific basis for many "high-tech" industries, ranging from the manufacture of computer chips to the fabrication of the tiny lasers that are used in Blu-ray players. The age of information technology could never have come about without the underpinning science of quantum physics.

Fifty years ago, computers were rare objects, and they were vast in size owing to the bulky components from which they were built. Nowadays, there are computers all around us. Not just desktop or laptop PCs and tablets, but mobile phones, televisions, car engines, and even washing machines each may contain computers. Modern computers are so compact because their circuitry comprises integrated circuits, sometimes also known as silicon chips. A miniaturized electronic circuit such as a computer CPU chip is only about 200 mm^2 in size yet contains the equivalent of around a million transistors per square millimetre. At the heart of every miniature component lie semiconductor devices. These act as tiny electronic switches, controlling the flow of electric current according to the fundamental rules of quantum physics, which determine the arrangement of electrons in the semiconductor materials.

Looking ahead, it is likely that the next decades will see advances in the area of quantum computing. Such devices rely on the inherent uncertainty that lies at the heart of quantum physics. Whereas a regular computer bit may be characterized as containing merely 0 or 1, a quantum qubit can exist in a superposition of states that is both 0 and 1 at the same time (just like Schrödinger's cat). As a result, quantum computers will be able to solve certain calculations much faster than conventional computers. They will also provide a means of transferring encoded information in an unbreakable format, a field referred to as quantum cryptography.

Many people believe that the digital revolution brought about by the availability of integrated circuits was one of the most significant technological advances in the entire history of the human race, comparable to the development of the wheel, the printing press, or the steam engine. It's worth remembering that it is an understanding of the underlying principles of quantum physics that allowed this revolution to occur.

5 Atoms

Having examined some of the distinctive characteristics of quantum physics, it is now time to apply these principles to learn about the structure of atoms, the processes in which their nuclei can participate, and the fundamental particles of which they are composed. As you read through Chapters 5, 6, and 7 you will see the ideas that have been developed in Chapters 3 and 4 being applied on smaller and smaller length scales. The evidence indicates that the principles of quantum physics, and the characteristic quantum phenomena of energy levels and indeterminacy, continue to provide a reliable guide to the nature of the physical world on all these length scales.

5.1 ATOMS AND IONS WITH ONE ELECTRON

The aim of this chapter is to develop an understanding of the structure of atoms in general. A start has already been made on this, in Chapters 3 and 4, by examining the hydrogen atom. Knowledge of the quantum behaviour of the one electron that is found in a hydrogen atom will provide a guide towards understanding the behaviour of the many electrons that are found in more complicated atoms. Of course, you should not expect to understand the full richness of atomic behaviour on the basis of a brief introduction to the hydrogen atom, but what you already know can be used to provide an insight into the behaviour of helium and lithium atoms (the next two simplest atoms after hydrogen), and those atoms illustrate many of the general principles that govern the structure of even more complicated atoms. This in turn is the basis for the chemical behaviour of atoms, which relies crucially on the way that the electrons are arranged around the nucleus in each different element.

If a neutral helium atom has one of its two electrons removed, it becomes a positively charged helium ion: He^+. Similarly, if two of the three electrons in a neutral lithium atom are removed, it becomes a positively charged lithium ion: Li^{2+}. Each of these ions is referred to as a hydrogen-like ion, because it contains a *single* electron. Now, as you know, the number of protons in the nucleus of any ion is just given by the atomic number (Z) of the element in question. It turns out that the energy levels of a hydrogen-like ion are each just Z^2 times larger than those of the hydrogen atom (Figure 5.1). Because $Z = 1$ for the hydrogen atom, the two equations are identical in that case.

In the case of the helium ion, He^+, $Z = 2$, so this implies that each energy level is four (i.e. Z^2) times larger than the corresponding value for the hydrogen atom. It also follows that the separation between any two energy levels in the He^+ ion will be four times greater than the separation between the corresponding levels in the hydrogen atom. Similarly, in the case of the doubly ionized lithium ion, Li^{2+}, $Z = 3$, the rule implies that each energy level is nine (i.e. Z^2) times larger than the corresponding value for the hydrogen atom. It also follows that the separation between

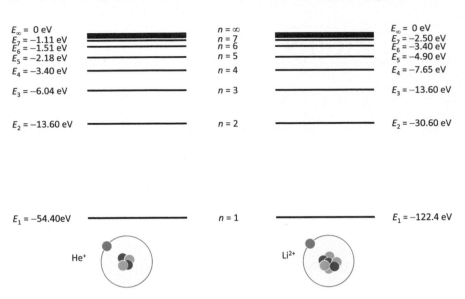

FIGURE 5.1   Hydrogen-like ions and their energy levels.

any two energy levels in the Li^{2+} ion will be nine times greater than the separation between the corresponding levels in the hydrogen atom. The photons absorbed or emitted by these ions therefore carry four or nine times the energy, respectively, of the photons involved in the corresponding transitions in hydrogen atoms. Transitions corresponding to the Balmer series in hydrogen will therefore give rise to absorption or emission lines in the UV part of the spectrum for He^+ and Li^{2+}, rather than in the optical part of the spectrum.

5.2   ATOMS AND IONS WITH TWO ELECTRONS

There is *no* simple formula for the energy levels of any neutral atom other than hydrogen. Quantum physics can predict the energy levels of the helium atom (for which $Z = 2$) to high precision, but it takes a lot of computer time to calculate this. It takes even more computer time to determine the energy levels for atoms with three or more electrons.

Another point to note is that, for hydrogen-like atoms and ions, the energy of each state is essentially determined by just the principal quantum number n. All 2s and 2p excited states for instance have virtually the same energy level in hydrogen, as noted in Chapter 3. In atoms with two or more electrons, things are no longer that simple. Interactions between the electrons mean that the energy levels of each state depend on the other quantum numbers, such as l, as well as on n.

The reason for this increasing complexity is not particularly related to quantum physics. Similar complications would arise in attempts to use Newton's laws to predict the behaviour of the Solar System if it contained two planets that attracted each other as strongly as each is attracted to the Sun. In the case of the helium atom, the analogous problem is to find the positions and velocities for two electrons, in states of definite energy, when each electron is subject to the attractive influence of the nucleus *and* to the repulsive influence of the other electron. Just in case this does not already sound challenging enough, bear in mind that this is quantum physics, so the positions and velocities of both electrons will be indeterminate!

In the case of the hydrogen-like ions, it was possible to be quite prescriptive about the quantum state that's occupied by the single bound electron. If the ion is in its ground state, the electron occupies a 1s state; if the ion is in an excited state, then the electron might occupy a 2s state, or a 2p state, or any of the excited states that were described in Chapter 3. These states will not be exactly like those of the hydrogen atom, but for these simple ions the electrons can be fitted into "hydrogen-like" states because the only thing that changes is the charge on the nucleus.

Now, moving on to consider atoms or ions with two bound electrons, such as He or Li$^+$, the state of the atom can once again be described in terms of the behaviour of the electrons. For instance, an atom in a quantum state described by 1s1s would indicate that it contains two electrons each occupying something like a 1s state of a hydrogen-like atom. An atom in a quantum state described by 1s2s would indicate two electrons, with one of them occupying a 1s hydrogen-like state and the other a 2s hydrogen-like state. But what *are* the energy levels corresponding to these quantum states?

Consider first the energy-level diagram for the helium atom, which has two electrons. The ground state of helium will be described by 1s1s and so will correspond to two bound electrons each occupying something like a 1s state of a hydrogen-like atom. To depict the energy-level diagram, the first thing to decide is where to position the zero of energy. In the case of the hydrogen atom, the zero was chosen to correspond to a state in which the electron just manages to escape from the nucleus. In the energy-level diagram for the helium atom, the zero energy level is chosen to correspond to a helium nucleus and *two* free electrons.

The energy level corresponding to the ground state of singly ionized helium He$^+$ is four times that of a hydrogen atom, namely -54.40 eV. Now suppose, for a moment, that the repulsion of the two electrons in the ground state of the helium atom could be ignored. Then the ground-state energy would correspond to assigning principal quantum number $n = 1$ to each electron, and adding the energies associated with each. This hypothetical ground state of a helium atom with two bound electrons might be expected to be at an energy of $2 \times (-54.40 \text{ eV}) = -108.8$ eV. In reality, the true ground state of the helium atom is significantly higher, at an energy of -78.98 eV.

To understand, qualitatively, why the true ground-state energy is higher than the hypothetical ground state with no interaction, think about where the two electrons may be found. The positions are indeterminate, but there is some probability of finding one electron far from the nucleus and one electron much closer to it. In this case, the distant electron effectively experiences only the electric force due to the He$^+$ ion, with a net electrical charge of one proton. Conversely, the electron that's much closer to the nucleus experiences the effect of the full nuclear charge of two protons. This being the quantum world, all possibilities between these two extremes are allowed, and the typical charge experienced by each electron will be something between one and two positive charges. The name given to this effect is screening, and you can think of it as one electron neutralizing, or cancelling out, part of the charge of the nucleus. You have already seen that the energies of the quantum states depended strongly on the nuclear charge. So, it should come as no surprise that, when one electron reduces the effective charge experienced by the other, the energy levels will change dramatically.

As noted, the ground-state energy of the helium *atom* is −78.98 eV and the ground-state energy of the helium *ion* is −54.40 eV. In order to turn an atom into an ion, the atom must be ionized and the ionization energy of the helium atom is the energy required to remove one electron from the atom in its ground state. The energy required to do this corresponds to the energy of the helium ion ground state minus the energy of the helium atom ground state: −54.40 eV − (−78.98 eV) = 24.58 eV. This turns out to be the *biggest* ionization energy for any neutral atom.

5.3   ATOMS WITH THREE OR MORE ELECTRONS

Increasing the charge on the helium nucleus by one more proton produces lithium. It takes about 122 eV to remove the single electron from a Li^{2+} ion (i.e. nine times the ionization energy of a hydrogen atom), and it takes about 74 eV to remove one of the two electrons from a Li^+ ion. However, it turns out that only about 5 eV is required to remove one of the three electrons from a neutral Li atom. No singly ionized ion requires more energy than Li^+ to make it doubly ionized, yet the lithium atom itself is rather easily ionized, much more easily than the helium atom or the hydrogen atom.

The key to understanding why is to consider what happens when a third electron is added to the ground state of Li^+, to make a neutral lithium atom. As you saw in Chapter 3, there are only two quantum states that have quantum numbers $n = 1$ and $l = 0$, i.e. there are only two 1s states, each of them has $m_l = 0$, one of them has $m_s = +1/2$, and the other has $m_s = -1/2$. You can think of these hydrogen-like states as though they can each be "occupied" by a single electron. So, the ground state of lithium *cannot* correspond to a quantum state of 1s1s1s.

The principle that bans the third electron from being in a similar state to the other two is a crucial result of quantum physics; it was suggested by Wolfgang Pauli in 1925. The Pauli exclusion principle bans any two electrons in the same atom from occupying the same quantum state.

Remember that for any value of n, there are two "s" states, six "p" states, ten "d" states, and so on. Each quantum state corresponds to a different allowed combination of the four quantum numbers. According to the exclusion principle, each of these quantum states can accommodate only one electron.

Because there are only two 1s states, the third electron in a lithium atom must occupy a 2s state. The ground state of the lithium atom can therefore be represented as 1s1s2s. The third electron in the lithium atom has principal quantum number $n = 2$, which makes it much more remote from the nucleus. As a result, one of the three electrons in the lithium atom is rather weakly bound, because it experiences a net charge that is not much greater than the one unit of the Li^+ ion. The other two units of nuclear charge are effectively screened by the other two more tightly bound electrons. A rough estimate of the ionization energy of the lithium atom would be somewhat greater than that for the 2s or 2p state of hydrogen, namely: 13.60 eV$/2^2 = 3.40$ eV. In fact, it is 5.39 eV, indicating that the other two electrons do not completely screen two units of nuclear charge.

It is a remarkable fact that lithium is a highly reactive metal, but helium is an extremely inert gas, yet the only difference between them is that lithium atoms contain three electrons surrounding a nucleus containing three protons, whereas helium atoms have two electrons surrounding a nucleus with two protons. The difference in properties is all due to the ionization energy difference between the two atoms, which in turn is due to the fact that quantum physics and the Pauli exclusion principle place a limit on the number of electrons in different quantum states. In the final section of this chapter, these ideas are taken even further to explore the basis of chemistry.

5.4 THE PERIODIC TABLE OF THE ELEMENTS

As you have seen, for simple atoms such as He and Li, it is possible to be quite prescriptive about the states occupied by the two or three electrons that now have to be considered. For the helium atom, both electrons can occupy 1s states, so the ground state of helium is 1s1s. For the lithium atom, you have seen that the ground state is 1s1s2s.

With more electrons, this notation soon gets unwieldy. To avoid this problem, a more compact way of describing the organization of the electrons around a nucleus is used, referred to as the electron configuration. For the helium atom with a ground state of 1s1s, the electron configuration is written as $1s^2$, and for the lithium atom with a ground state of 1s1s2s, the electron configuration is written as $1s^2\,2s^1$. It is important to realize that the superscripts here *do not* refer to "powers" of numbers, they are merely labels indicating the number of electrons that occupy the states specified by the n and l quantum numbers.

It is often quite convenient to think of electrons as "filling up" successive subshells in the atom. For instance: the 1s subshell can accommodate two electrons because there are two 1s quantum states, the 2s subshell can accommodate two more electrons, and the 2p subshell can accommodate a further six electrons because there are six 2p quantum states. Subshells are filled in order of increasing energy so that the atom as a whole has the lowest possible energy level. The start of the sequence of filling states in order of increasing energy turns out to be 1s, 2s, 2p, 3s, 3p, 4s, 3d, 4p, 5s, 4d, 5p ... which gets as far as atomic number 54 (corresponding to xenon). Although such terminology is not entirely accurate, owing to the interactions between electrons, it provides a useful simplification. For atoms that contain several electrons the states are somewhat different from the "hydrogen-like" quantum states, because the interactions between the electrons complicate matters.

This can be seen as the origin of the wonderful structure encapsulated by the periodic table of the elements, first laid out by the Russian chemist Dmitri Mendeleev around 1869 (Figure 5.2). He arranged the various elements in a grid such that

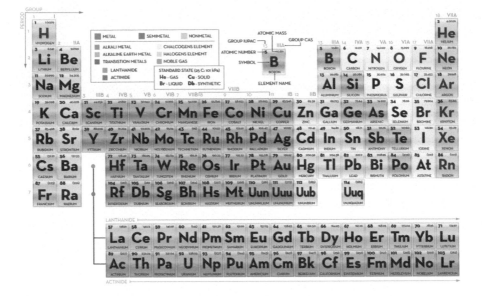

FIGURE 5.2   The periodic table of the elements.

elements with similar properties appeared in the same column (or Group) of the table, and each row (or Period) of the table is now recognized as containing the elements in order of increasing atomic number.

Lithium (Li), sodium (Na), and potassium (K), known as the alkali metals, each lie in the first Group of the table, because they are each similarly very reactive. The outer electron in each of these elements is a solitary electron occupying an "s" subshell. The electron configurations of these three elements are: Li: $1s^2 2s^1$, Na: $1s^2 2s^2 2p^6 3s^1$, and K: $1s^2 2s^2 2p^6 3s^2 3p^6 4s^1$, corresponding to atomic numbers of 3, 11, and 19. That single outer electron is what makes each of these elements so chemically reactive, since it is not tightly bound and so can readily exchange with other elements.

The right-most Group of the table contains the inert (or so-called noble) gases helium (He), neon (Ne), argon (Ar), and krypton (Kr). The outermost electrons of each of these elements comprise a full subshell. The electron configurations of these four elements are: He: $1s^2$, Ne: $1s^2 2s^2 2p^6$, Ar: $1s^2 2s^2 2p^6 3s^2 3p^6$, and Kr: $1s^2 2s^2 2p^6 3s^2 3p^6 3d^{10} 4s^2 4p^6$, corresponding to atomic numbers of 2, 10, 18, and 36. That full outer subshell is what makes each of these elements so chemically unreactive, since there are no loosely bound electrons to share with other elements in chemical reactions.

As a final example, the penultimate Group on the right-hand side of the table contains a group of elements known as the halogens, which include fluorine (F), chlorine (Cl), and bromine (Br). The outermost electrons of each of these elements comprise a subshell with a single vacancy. The electron configurations of these three elements are: F: $1s^2\ 2s^2\ 2p^5$, Cl: $1s^2\ 2s^2\ 2p^6\ 3s^2\ 3p^5$, and Br: $1s^2\ 2s^2\ 2p^6\ 3s^2\ 3p^6\ 3d^{10}\ 4s^2$ $4p^5$, corresponding to atomic numbers of 9, 17, and 35. The vacancy in the outer sub-shell means that these elements readily accept an electron from another atom (such as those in the first Group) to form strongly bonded compounds. This is why sodium chloride (NaCl), whose crystals are formed from sodium and chlorine atoms bound together, is such a widespread substance on Earth, known as common salt.

All the richness of chemistry and chemical reactions essentially stems from these simple quantum rules for how the electrons are distributed in atoms. Although this section of the book focusses on the smallest scale structures of the Universe, the behaviour of materials on an everyday scale that we see around us is firmly rooted in this quantum world. The behaviour of solids, liquids, and gases, the differences between metals, organic compounds, and other materials are all determined by the simple rules that operate at an atomic scale.

6 Nuclei

The next leg of the journey is to travel further inside the atom to look at the quantum world of nuclei. You know already that nuclei have two building blocks: neutrons and protons, and these are referred to collectively as nucleons. Different combinations of nucleons account for the different isotopes in the world around us.

Even though you will be entering a new realm, you can be sure that there are many similarities between the quantum world of atoms and electrons that you have just read about, and the quantum world of nuclei. For instance, as in the case of an atom, only certain energies are allowed to a nucleus. Quantum jumps occur between the energy levels, with the emission or absorption of photons. In the case of nuclei, these are gamma-ray photons, which have energies around 1 MeV (remember 1 MeV = 10^6 eV), i.e. several hundred thousand times greater than those of a visible photon (with energies of just a few eV). Another similarity is that, as in the case of electrons in atoms, the positions and velocities of nucleons in nuclei are subject to indeterminacy. The distinctive feature of nuclei is that the nucleons are confined in a volume that is less than 10^{-14} m in diameter – more than 10^4 times smaller than the diameter of an atom.

However, the possibilities for changes in nuclei are far more varied than those open to electrons in atoms. The big difference is that nuclei can also *transform* from one element to another in a variety of nuclear decays and nuclear reactions. You may know that rocks and fossils may be dated by measuring the relative proportions of radioactive parent and daughter nuclei in certain minerals. This chapter examines how these radioactive transformations occur.

Nuclear decays are spontaneous changes that occur in radioactive isotopes, causing the individual nuclei to transform from one type to another. Photons and particles may both be emitted as a result of these decays, carrying away energy in the process. Nuclear decay processes have been known about since the end of the 19th century, from the work of scientists such as Henri Becquerel, Ernest Rutherford, Marie Sklodowska Curie, and Pierre Curie. This chapter discusses the characteristic features of all radioactive decays, before focusing in turn on the three types of radioactive decay process that these eminent scientists discovered: alpha-decay (α-decay), beta-decay (β-decay), and gamma-decay (γ-decay).

Nuclear reactions refer more generally to processes in which two or more nuclei are involved, again resulting in the creation of different nuclei from those originally present. In particular, the latter part of this chapter focuses on nuclear fission (in which nuclei are split apart) and nuclear fusion (in which nuclei are joined together), releasing energy in each case.

6.1   MASS AND ENERGY

You've already met the idea of the conservation of energy and seen that it arises due to the symmetry of physical laws with respect to time. Another conservation law is that of the conservation of electrical charge. In a similar way to the other conserva-tion laws that were linked to symmetries by Emmy Noether, charge conservation is related to the fact that the electric and magnetic fields are not changed by different choices of the value representing the zero point of electrostatic potential energy.

As you will see later, in all nuclear processes the following conservation rules are obeyed:

- *Electric charge* is conserved: the net charge of the products of a nuclear process is the same as the net charge of the original nucleus or nuclei.
- *Mass number* is conserved: the total number of nucleons in the products is the same as that in the original nucleus or nuclei.
- *Energy* is conserved (as long as mass is considered as a form of energy).

Because the energies involved in nuclear processes are so large, the relationship between energy and mass, $E = mc^2$, has to be taken into account. Remember from Chapter 2 that it is often convenient to express masses in units of "energy/c^2," such as MeV/c^2.

The question arises: *why* do certain nuclei transform from one type to another? As you might guess, it is all to do with energy (or equivalently, mass). The first thing to note is that the mass of *any* nucleus is *less than* the total masses of the individual protons and neutrons of which it is composed. For instance, a helium-4 nucleus has a mass of 3727.4 MeV/c^2, whereas the (precise) masses of the two protons and two neutrons of which it is composed have a total mass individually of:

$$(2 \times 938.28)\ \text{MeV}/c^2 + (2 \times 939.57)\ \text{MeV}/c^2 = 3755.7\ \text{MeV}/c^2.$$

The mass difference of:

$$(3755.7 - 3727.4)\ \text{MeV}/c^2 = 28.3\ \text{MeV}/c^2$$

is referred to as the mass defect of the helium-4 nucleus. It is the amount by which the nucleus is less massive than its constituent nucleons. Similarly, the binding energy of the helium-4 nucleus is −28.3 MeV (note the minus sign). In forming the helium-4 nucleus from its constituent nucleons, 28.3 MeV of energy is released, as a result of the mass decrease.

Conversely, in order to split the helium-4 nucleus apart, 28.3 MeV of energy must be supplied to create the extra mass of the constituent nucleons. This is a similar concept to the ionization energy that must be supplied to an atom in order to separate an electron from the rest of the atom.

Different nuclei have different mass defects and therefore different amounts of binding energy. Those with the *highest* mass defect have the *lowest* (i.e. most

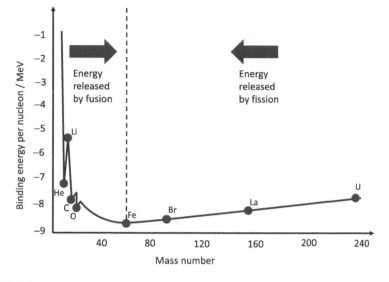

FIGURE 6.1   The binding energy per nucleon.

negative) binding energy – both characterized by the same numerical value, as indicated above for helium-4. As you might guess, the binding energy of heavy nuclei is generally lower (i.e. more negative) than the binding energy of light nuclei, simply because they have more nucleons to separate. However, a useful quantity is the binding energy per nucleon, calculated simply as the binding energy of the nucleus divided by the number of nucleons it contains. This quantity indicates how strongly each nucleon is bound to the nucleus. For helium-4, which is composed of four nucleons, the binding energy per nucleon is -28.3 MeV/4 = -7.1 MeV (Figure 6.1).

The isotope with the lowest (i.e. most negative) binding energy per nucleon, and therefore the most stable to decay, is iron-56, which has a binding energy per nucleon of -8.8 MeV. The most massive stable isotope is bismuth-209; it has the lowest binding energy of any stable nuclide, with a value of -1640 MeV. On a graph of the binding energy per nucleon versus atomic number, iron-56 sits at the lowest point of the curve, which turns up at both ends.

6.2   ALPHA-DECAY

An alpha-particle (α-particle) is the same as a helium nucleus, $^4_2$He, with mass number $A = 4$ and atomic number $Z = 2$. It consists of two protons and two neutrons. As noted above, it is a very tightly bound arrangement: this ground state for two protons and two neutrons has an energy that is 28.3 MeV lower than the energy of the four free nucleons of which it is composed.

In some cases, it is energetically favourable for a nucleus of mass number A and atomic number Z to move to a lower binding energy by emitting an alpha-particle and thereby producing a new nucleus, with mass number $A - 4$ and atomic number

$Z - 2$. A case in point is the unstable isotope of uranium, $^{234}_{92}U$, which contains 92 protons and 142 neutrons. This nucleus undergoes alpha-decay to produce an isotope of thorium, with 90 protons and 140 neutrons:

$$^{234}_{92}U \rightarrow {}^{230}_{90}Th + {}^{4}_{2}He$$

Notice that the values for A and Z are the same on the left-hand side of this equation (234 and 92 respectively) as the totals for A and Z on the right-hand side ($230 + 4$ and $90 + 2$ respectively), which bears out the assertions earlier that electric charge and mass number are both conserved in nuclear decays.

Both electric charge and mass number are conserved in an alpha-decay process because there is the same number of protons and neutrons in the products as in the original nucleus. But what about energy conservation?

In the example above, the binding energy of the uranium-234 nucleus is -1778.6 MeV, while the binding energies of the thorium-230 and helium-4 nuclei are -1755.1 MeV and -28.3 MeV respectively. The total binding energy of the products is therefore -1783.4 MeV. The reaction involves moving to a configuration in which the binding energy has *decreased* by $(-1778.6 \text{ MeV}) - (-1783.4 \text{ MeV}) = 4.8$ MeV. So in this case, the alpha-decay of the uranium-234 nucleus liberates 4.8 MeV of energy. This takes the form of kinetic energy and is carried away (almost exclusively) by the alpha-particle. The decay can be written as

$$^{234}_{92}U \rightarrow {}^{230}_{90}Th + {}^{4}_{2}He + 4.8 \text{ MeV}$$

It can therefore also be said that the combined mass of the thorium-230 and helium-4 nuclei is 4.8 MeV/c^2 less than the mass of the original uranium-234 nucleus.

6.3   BETA-DECAY

This section looks at the second mode of nuclear decay, namely beta-decay. In fact, there are three related processes, each of which is a type of beta-decay. The first one to consider is beta-minus (β^-) decay which involves the emission of an *electron* from the nucleus of an atom. Perhaps the first question to address is: where did that electron come from? Was it there in the first place? The answer is no; it is quite impossible for nuclei to "contain" electrons. The electron is *created*, by the decay, just as a photon is created when an atom makes a transition from a higher energy level to a lower energy level.

In order to see how the electron comes into existence, consider a specific case, namely the beta-minus decay of the lead isotope $^{214}_{82}Pb$. The electric charge of the lead nucleus is that due to 82 protons, so is equal to 82 units of positive charge. The electron emitted by the nucleus carries one unit of negative charge. Because charge is conserved in all nuclear decay processes, the charge of the nucleus *before* the decay must be equal to the charge of the nucleus *after* the decay plus the charge of the electron that is emitted. As a result, the charge of the resultant nucleus is 83 units of positive charge, and the resultant nucleus must therefore contain 83 protons; it is a nucleus of the element bismuth.

So the resultant bismuth nucleus has one more proton than the original lead nucleus. But where has that proton come from? To answer that you need to consider the mass number of the nucleus. Because mass number is also conserved in nuclear decays, the total number of nucleons (i.e. protons plus neutrons) in the resultant bismuth nucleus must be the same as in the original lead nucleus. It therefore contains 214 nucleons and so the resultant nucleus is the bismuth isotope $^{214}_{83}\text{Bi}$.

Because the bismuth nucleus contains one more proton than the lead nucleus, but the same number of nucleons in total, it must contain one less neutron than the original lead nucleus. What has happened is that one of the neutrons in the lead nucleus has transformed into a proton, with the emission of an electron. Transformations between neutrons and protons lie at the heart of all beta-decay processes.

For reasons that will become clearer in Chapter 12, another particle is created in the beta-decay process too. It is called the electron antineutrino and it has zero electric charge. An electron and an electron antineutrino are always created in a beta-minus decay, the minus sign indicating that the electron is negatively charged. The existence of neutrinos was predicted by Wolfgang Pauli in 1930 to explain how beta-decays were able to conserve energy and momentum, but neutrinos were not experimentally detected until 1956. The overall decay process for this lead isotope is therefore:

$$^{214}_{82}\text{Pb} \rightarrow {}^{214}_{83}\text{Bi} + e^- + \bar{v}_e$$

The rather clumsy symbol $\bar{v}_e$ represents an electron antineutrino, where v is the Greek letter *nu* (pronounced "new"). The subscript "e" indicates that it is associated with an electron, and the bar over the top of the letter indicates that it is an antimatter particle. As in any nuclear decay, the mass of the products is less than the mass of the original nucleus, and the difference in mass is liberated as kinetic energy of the products.

The process described above is only part of the story as far as beta-decay is concerned. There is a very closely related process, called beta-plus (β^+) decay, in which a positron (i.e. an antielectron) is created, along with an electron neutrino, which has zero charge. In this process, a proton in the original nucleus transforms into a neutron, so *decreasing* the atomic number by one. A nucleus that undergoes beta-plus decay is the unstable oxygen isotope $^{14}_8\text{O}$ which transforms into a stable nitrogen isotope $^{14}_7\text{N}$. The decay in this case can be represented as:

$$^{14}_8\text{O} \rightarrow {}^{14}_7\text{N} + e^+ + v_e$$

Here, the symbol e^+ is used to represent the positron and v_e is the electron neutrino.

The final process in the suite of beta-decays is that of electron capture. As suggested by its name, in this process a nucleus captures an electron, usually from the inner regions of the electron cloud surrounding it. A proton in the nucleus interacts with the captured electron, forming a neutron and emitting an electron neutrino. As in beta-plus decay, a proton in the original nucleus transforms into a neutron, so *decreasing* the atomic number by one. A nucleus that undergoes electron capture is the unstable aluminium isotope $^{26}_{13}$Al, which transforms into a stable magnesium isotope $^{26}_{12}$Mg. The decay in this case can be represented as:

$$^{26}_{13}\text{Al} + e^- \rightarrow {}^{26}_{12}\text{Mg} + \nu_e$$

6.4 GAMMA-DECAY

The final type of nuclear decay considered here is gamma-decay. In contrast to the two processes of alpha- and beta-decay, this involves no change in the numbers of neutrons and protons. Gamma-decay occurs when a nucleus finds itself in an excited state. A quantum jump down to a lower-energy quantum state, with the same number of neutrons and protons, is accompanied by the emission of a photon, as with transitions in atoms. This time, however, the photon energy is around a million time larger – it is a gamma-ray photon. Such excited states of nuclei may be created as a result of alpha-decay or beta-decay processes, or by the collisions of nuclei at high kinetic energies.

As an example, consider the unstable isotope of caesium, $^{137}_{55}$Cs, which undergoes beta-minus decay to produce an excited state of the barium isotope, $^{137}_{56}$Ba. The barium nucleus then decays to its ground state with the emission of a gamma-ray photon of energy 662 keV. In the process of gamma-decay, the number of protons and neutrons in the nucleus remains unchanged. So the atomic number and mass number of the barium nucleus after the gamma-decay are the same as they were before, namely 56 and 137 respectively.

6.5 HALF-LIFE

A wonderful example of the quantum indeterminacy that lies at the heart of matter is to be found in the idea of radioactive half-life. When observing a sample of radioactive material undergoing either alpha-decay, beta-decay, or gamma-decay, if a graph is plotted of the number of decays per second against time, it would show a characteristic shape: a curve which continually decreases at an ever-shallower rate. In fact, the graph would fall by a factor of two for every half-life that elapses. The half-life is therefore the time for half of a sample of radioactive nuclei to decay. Each radioactive isotope has its own characteristic half-life: some may be as short as the tiniest fraction of a second; others, such as uranium-238, have half-lives of billions of years.

As an example, the naturally occurring radioactive isotope carbon-14 undergoes beta-minus decay (to nitrogen-14) with a half-life of 5730 years. The proportion of carbon-14 to (normal) carbon-12 in living things is well known, therefore by examining the remaining proportion of carbon-14 in things long dead (the wooden timbers in ancient buildings for instance) the amount of time since the item ceased to live may be calculated. There are complications to do with the varying proportion of carbon-14 in the Earth's atmosphere over recent history, and contamination of samples by more recent biological matter, but the principle is sound. The technique is useful for ages up to about ten half-lives (around 60,000 years in the case of carbon-14). For objects older than this, less than one-thousandth of the original carbon-14 remains, and the accuracy of the measurement is unreliable.

The constancy of the half-life for a given isotope shows that the probability for a given nucleus to decay (in, say, the next second) remains constant. So these decay processes are intrinsically unpredictable, event-by-event, yet follow a wonderfully simple rule when studying large numbers of decays: the radioactivity is proportional to the number of radioactive nuclei that remain. Nuclei have no past: they never grow old; they either stay exactly the same or change dramatically.

6.6 NUCLEAR FISSION

Nuclear fission is the name given to the process in which a relatively massive nucleus splits apart into two less massive nuclei, of roughly equal size. In some types of nuclei, this process may occur spontaneously, while in others it may be induced by bombarding a nucleus with other particles. An example will serve to illustrate the principles involved.

If a nucleus of the isotope uranium-235 absorbs a slow-moving neutron, it will temporarily form a nucleus of uranium-236. This nucleus is unstable and so will rapidly split apart to form two smaller nuclei, such as krypton-92 and barium-141, plus a few "left-over" neutrons. The particular nuclei, and specific number of neutrons, formed by each such interaction are random, but these fission products are usually themselves radioactive, and subsequently decay via a range of beta- and gamma-decays. In the case stated above, the reaction may be written as:

$$^{235}_{92}\text{U} + ^{1}_{0}\text{n} \rightarrow ^{236}_{92}\text{U} \rightarrow ^{92}_{36}\text{Kr} + ^{141}_{56}\text{Ba} + 3^{1}_{0}\text{n}$$

Note that here, *three* neutrons are ejected from the fission process, and a neutron can be written as "$^{1}_{0}\text{n}$" to emphasize that it has a charge of zero, but a mass number of one.

Now look at the binding energies of the participants in this reaction. The binding energy of uranium-235 is −1783.9 MeV, while the binding energies of the krypton-92 and barium-141 nuclei are −783.2 MeV and −1174.0 MeV respectively. The total binding energy of the product nuclei is therefore −1957.2 MeV, and this is significantly lower than that of the original nucleus. The difference in energy, 173.3 MeV, is released principally as the kinetic energy of the free neutrons. If these free neutrons are slowed down, they are able to interact with any further uranium-235 nuclei present in the sample, and so initiate a self-sustaining chain reaction.

Such reactions have been employed both in nuclear power stations, where the energy released is used to heat and vaporize water that turns turbines to generate electricity, and in nuclear weapons, where the energy released is put to more destructive use. But such reactions also occur naturally in the Earth. There is a site at Oklo in West Africa where the concentrations of various naturally occurring uranium isotopes indicate that a self-sustaining chain reaction occurred here about 1.7 billion years ago. In this natural reactor, it is believed that water filtering through crevices in the rock played a key role. Just as in a nuclear power station, the water slowed down the neutrons that were emitted during the fission process so that they could inter-act with other uranium nuclei. Without the water to slow them down, the neutrons would move so fast that they would just bounce off and not produce a chain reaction. When the rate of heat production from the reactions became so high that the water vaporized, the neutrons were not slowed down and the reactions stopped until the water cooled again. At this point the process could begin again. The chain reactions stopped completely when the abundance of uranium-235 nuclei became too low to keep the reactions going.

6.7   NUCLEAR FUSION

Nuclear fusion is the name given to the process by which two (or more) low-mass nuclei join together to form a heavier nucleus. In general, these reactions only occur at extremely high temperatures, because the positively charged nuclei need to be "forced" close enough together in order for the fusion to happen. At low temperatures, the electrical repulsion between them keeps them far enough apart not to interact. Like nuclear fission, nuclear fusion too is a natural process that occurs throughout the Universe. In fact, it is the process that powers stars, including the Sun, and is largely responsible for creating the range of elements that make up the world around us. You will look at the consequences of fusion reactions in the Universe more closely in Chapter 16, but first consider a particular set of reactions – those occurring in the Sun and other low-mass stars – which illustrates the principles.

In 1920 the British astronomer Arthur Eddington had first proposed that the Sun and other stars are powered by nuclear fusion. In fact, energy is generated in the core of the Sun by a set of nuclear fusion reactions known as the proton–proton chain (or pp chain for short), the details of which were first worked out by Hans Bethe and Charles Crichtfield in 1938 (Figure 6.2). The chain begins with two protons (hydro-gen nuclei) fusing together to make a nucleus of deuterium (heavy hydrogen).

$$^1_1\text{H} + ^1_1\text{H} \rightarrow ^2_1\text{H} + \text{e}^+ + \nu_\text{e} \quad \text{(step 1)}$$

The net effect of this reaction is that one of the original protons has transformed into a neutron with the emission of a positron and an electron neutrino. This is therefore similar to the process of beta-plus decay which you met in Section 6.3. Because of the conversion of a proton into a neutron, it cannot just be assumed that the binding energy of the deuterium nucleus (–2.2 MeV) is released here. Instead note that the reactants have a mass-energy of $(2 \times 938.3 \text{ MeV}) = 1876.6 \text{ MeV}$, and the products have a mass-energy of 1875.7 MeV (for the deuterium) + 0.5 MeV (for the positron) plus a negligible amount for the electron neutrino, i.e. a total of 1876.2 MeV. So, the amount of energy liberated is (1876.6 MeV – 1876.2 MeV) =

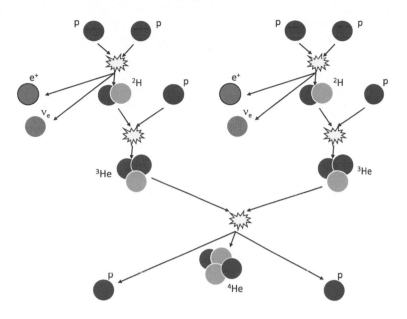

FIGURE 6.2   The proton–proton chain.

0.4 MeV. In addition, the positron will immediately annihilate with an electron to give a further 2×0.5 MeV = 1.0 MeV in the form of photons. So the total energy release from this reaction is (0.4 MeV + 1.0 MeV) = 1.4 MeV.

The next step in the chain is that the deuterium nucleus quickly captures another proton (hydrogen nucleus), to make a nucleus of helium-3:

$$^2_1\text{H} + ^1_1\text{H} \rightarrow ^3_2\text{He} \quad \text{(step 2)}$$

Here, the binding energy of the reactants is –2.2 MeV (deuterium) + 0 MeV (hydrogen), and that of the product is –7.7 MeV (helium-3). So a further 5.5 MeV of energy is released in this step.

The final step is that two of these helium-3 nuclei fuse together to form a nucleus of helium-4, releasing two protons (hydrogen nuclei) back into the mix:

$$^3_2\text{He} + ^3_2\text{He} \rightarrow ^4_2\text{He} + ^1_1\text{H} + ^1_1\text{H} \quad \text{(step 3)}$$

Here, the binding energy of helium-3 is –7.7 MeV and the binding energy of helium-4 is –28.3 MeV. So, the binding energy of the reactants is (–7.7 MeV) + (–7.7 MeV) = –15.4 MeV, while that of the products is (–28.3 MeV) + 0 MeV. Therefore an energy of (–15.4 MeV) – (–28.3 MeV) = 12.9 MeV is released in the final step.

Now, two instances of each of the first and second steps in the chain are needed for each instance of the third step in the chain (in order to make the two helium-3 nuclei required), so the complete reaction is that six protons have been combined to make a helium-4 nucleus plus two protons, two positrons, and two electron neutrinos:

$$6\,^1_1\text{H} \rightarrow\,^4_2\text{He} + 2\,^1_1\text{H} + 2e^+ + 2\nu_e$$

The two positrons in turn will annihilate with two electrons, producing photons

$$2e^+ + 2e^- \rightarrow \text{photons}$$

The net effect is therefore that four protons and two electrons have been combined to make a single helium-4 nucleus and two electron neutrinos, plus some photons:

$$4\,^1_1\text{H} + 2e^- \rightarrow\,^4_2\text{He} + 2\nu_e + \text{photons}$$

The total energy release is (1.4 MeV × 2) from step 1, plus (5.5 MeV × 2) from step 2, plus 12.9 MeV from step 3. Adding these together gives 26.7 MeV. Notice that this is slightly less than the binding energy of the helium-4 nucleus (28.3 MeV), as it has been created from four protons rather than two protons and two neutrons.

In general, fusion reactions to build more massive nuclei from lighter ones will release energy all the way up to the nucleus iron-56, which, as was noted earlier, has the lowest (most negative) binding energy per nucleon. As you will see in Chapter 17, this proves to be crucial in determining the life cycles of stars.

7 Particles

Half a century ago, the account of subatomic structure would have ended with nucleons. But now a third layer of structure is known: it is believed that protons and neutrons are composed of structureless particles known as quarks. This chapter will be looking at an area that is at the forefront of scientific research today. The aim here is nothing less than an understanding of the fundamental constituents from which the Universe is built.

So far in this book, you have met four types of subatomic particle: electrons, electron neutrinos, protons, and neutrons, along with their antiparticles in some cases. Over the last 50 years or so, experiments have revealed that other subatomic particles exist, and that those mentioned above are merely the common representatives of two distinct classes of object, known as leptons and hadrons. These are now considered in turn in order to complete the final leg of your journey to take the world apart.

7.1 LEPTONS

Electrons (e^-) and electron neutrinos (ν_e), together with their antiparticles, are believed to be fundamental particles. By fundamental, is meant that there is no evidence that they are composed of smaller or simpler constituents. Furthermore, two more particles, with the same charge as the electron, only rather heavier, were discovered in 1936 and 1975. The first is known as the muon (represented by μ^-; the Greek letter *mu*, rhymes with "cue") which is about 200 times heavier than the electron. The second is called the tau (represented by τ^-; the Greek letter *tau*, rhymes with "cow") which is about 3500 times heavier than the electron. The superscript minus signs on the electron, muon, and tau indicate that these particles have negative electric charge. Muons and taus are unstable, and rapidly decay into their less massive sibling, the electron, in a fraction of a second. Like the electron, the muon and tau each have an associated neutrino: the muon neutrino (ν_μ) and the tau neutrino (ν_τ) with zero electric charge.

These six fundamental particles are collectively referred to as leptons. (The word lepton comes from the Greek *leptos*, meaning "thin" or "lightweight.") The six different types are often rather whimsically referred to as different flavours of lepton, and the three pairs of particles are often referred to as three generations of leptons (Figure 7.1).

To each lepton there corresponds an antilepton with opposite charge but with the same mass. These are denoted by the symbols e^+, μ^+, and τ^+ for the charged leptons and $\bar{\nu}_e$, $\bar{\nu}_\mu$, and $\bar{\nu}_\tau$ for the neutral leptons.

Until recently, it was not known whether the three types of neutrino possess mass, like the charged leptons, or whether they are massless, like photons. However, in

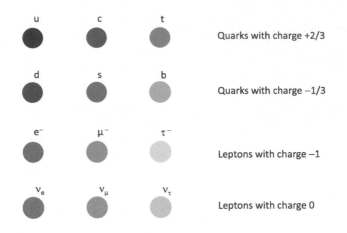

FIGURE 7.1   Three generations of leptons and quarks.

2005 results from the Canadian *Sudbury Neutrino Observatory*, studying the neutrinos emitted by the Sun, showed that they undergo flavour oscillations. Some of the electron neutrinos produced by the proton–proton chain in the core of the Sun *change* into muon neutrinos or tau neutrinos as they travel through space before reaching the Earth. This is the solution to the long-standing problem of why fewer electron neutrinos are detected from the Sun than are believed to be produced in its core. The implication of this is that neutrinos do indeed possess mass, as massless neutrinos would not be able to change flavour, according to current theories of how these particles behave. Measurements show that the *total* mass of the electron neutrino, muon neutrino, and tau neutrino must be less than 0.12 eV/c^2.

7.2   QUARKS

The other subatomic particles that you have met so far – protons and neutrons – are examples of hadrons. (The word hadron comes from the Greek *hadros*, meaning "strong" or "robust.") Although the only hadrons existing in the everyday world are protons and neutrons, many more types of hadron can be created in high-energy collisions of nucleons. Such reactions are common in the upper atmosphere, where high-energy protons from outer space (known as cosmic-ray protons) collide with nuclei of nitrogen and oxygen, smashing them apart and creating new hadrons. Since the 1960s, such reactions have been closely studied under controlled conditions, in high-energy physics laboratories, where protons and electrons are accelerated to high kinetic energies using very high voltages.

Although many dozens of different types of hadron may be created in this way, all of the new ones are unstable and they rapidly decay into other, long-lived particles, such as leptons, protons, and neutrons. Fortunately, it's not necessary to dwell on (let alone remember) the names and properties of all the types of hadron, because there is a straightforward description for building them from particles that *are* believed to be fundamental, namely from quarks and antiquarks, whose existence was first proposed by Murray Gell-Mann and George Zweig in 1964.

There are six flavours of quark, labelled by the letters u, d, c, s, t, and b, which stand for up, down, charm, strange, top, and bottom. (The last two are sometimes referred to as "truth" and "beauty" instead of "top" and "bottom.") The up, charm, and top quarks each carry a positive charge equal to 2/3 that of a proton, while the down, strange, and bottom quarks each carry a negative charge equal to 1/3 that of an electron. Like the leptons, the six quarks are often grouped into three generations, on the basis of their mass, with the first generation being the least massive. To each quark, there corresponds an antiquark, with the opposite charge and the same mass. These are denoted by the symbols $\bar{u}$, $\bar{d}$, $\bar{c}$, $\bar{s}$, $\bar{t}$, and $\bar{b}$. So anti-up, anti-charm, and anti-top antiquarks each carry a negative charge equal to 2/3 that of an electron, while anti-down, anti-strange, and anti-bottom antiquarks each carry a positive charge equal to 1/3 that of a proton.

The up quarks and down quarks are the constituents of protons and neutrons, and along with their antiquark counterparts are the least massive of all the quarks. The charm and strange quarks and antiquarks are more massive than the up and down quarks, and the top and bottom quarks and antiquarks are yet more massive still. The large masses of these second- and third-generation quarks are the reason why powerful particle accelerators are required to produce them. In order to create this amount of mass, a large amount of kinetic energy must be supplied in accordance with $E = mc^2$. In fact, hadrons containing top quarks were first detected in 1995; they have masses nearly 200 times that of the proton.

Quarks and antiquarks have *never* been observed in isolation. They only occur bound together inside hadrons. In fact, there are three confirmed recipes for building hadrons from quarks, as follows. A hadron can consist of:

- Three quarks (in which case it is called a baryon) or
- Three antiquarks (in which case it is called an antibaryon) or
- One quark and one antiquark (in which case it is called a meson)

Any combination of quarks and antiquarks that obeys one of these three recipes is a valid hadron, and the net electric charge of a hadron is simply the sum of the electric charges of the quarks or antiquarks of which it is composed (Figure 7.2). Notice that the net charge of a hadron is therefore *always* a whole number, despite the fact that the quarks themselves have non-whole number electric charge.

As a specific example of the hadron-building recipe, the proton is a baryon, so it is composed of three quarks, and as mentioned above, it is composed of up and down quarks only. Now, the proton has a positive charge and the only way that three up or down quarks can be combined to make this net charge is by combining two up quarks with a down quark. So the quark content of a proton is (uud), giving a net charge of +2/3 + 2/3 − 1/3 = +1. Likewise, because the neutron has charge zero and is composed of three up or down quarks, its quark content must be (udd), giving a net charge of +2/3 − 1/3 − 1/3 = 0.

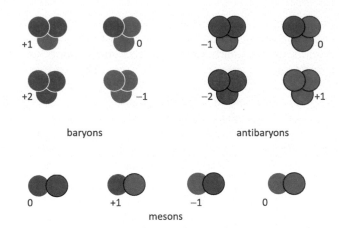

FIGURE 7.2   Baryons, antibaryons, and mesons with their electric charges.

In recent years there have been claims for the detection of particles dubbed tetraquarks and pentaquarks which, as you might expect, consist of four or five quarks or antiquarks (in fact they appear to contain either two quarks and two antiquarks or four quarks and one antiquark). Although claims for their existence are not yet verified, if they do exist, they only hang around for a tiny fraction of a second before decaying into more familiar baryons and mesons.

This tally of six leptons and six quarks, each with their own antiparticles (so 24 in all!), may seem like a huge number of fundamental particles. However, don't let this put you off. Virtually everything in the Universe is made up of merely the first generation of each type, namely electrons, up quarks, and down quarks, with electron neutrinos being created in beta-decays. The second generation of leptons (μ^- and ν_μ) and the second generation of quarks (c and s), the third generation of leptons (τ^- and ν_τ) and the third generation of quarks (t and b), have exactly the same properties as their first-generation counterparts except that they are more massive. Quite why nature decided to repeat this invention three times over is not currently understood, but you may be pleased to learn that current theories do not predict the existence of any more generations.

7.3   HIGH-ENERGY REACTIONS

In particle accelerators, or in the Earth's upper atmosphere, hadrons can smash into each other with large kinetic energies, and create new hadrons from the debris left behind.

As an example of the high-energy reactions that can occur, when a proton with kinetic energy of several hundred MeV collides with a nucleon, new hadrons, called pions, can be created. In such processes, some of the kinetic energy of the protons is converted into mass, via the familiar equation $E = mc^2$, and so appears in the form of new particles. The pions that are created come in three varieties: π^+, with positive charge (the same as a proton), π^- with negative charge (the same as an electron), and a neutral π^0 with zero charge (like a neutron). Their masses are each around 140 MeV/c^2, so 140 MeV of kinetic energy is required in order to create each pion. Pions are examples of mesons and so are composed of a quark and an antiquark. In fact, pions are the least massive mesons and (like protons and neutrons) are composed of up and down quarks. To get a positive charge on π^+, the combination (u$\bar{\text{d}}$) is needed, with charge +2/3 + 1/3 = +1. Likewise, the negatively charged pion, π^-, must have the combination (d$\bar{\text{u}}$), with charge −1/3 − 2/3 = −1.

But now there arises an interesting question about π^0: is it (u$\bar{\text{u}}$) or (d$\bar{\text{d}}$)? Each possibility follows the rules, and each produces zero charge. The general principle of quantum physics seems to be that "whatever is not strictly forbidden will be found to occur." Each possibility is allowed; each is found to occur. But that does not mean that are there two types of π^0. It means that if we carry out a suitable experiment to determine the content of π^0, sometimes one answer will be found, sometimes the other. You have seen similar things before in the quantum world. For instance, in atomic physics, it is only the act of measurement that forces one of the possible outcomes for the position of an electron. In particle physics, the very notion of what something is made of is subject to the same indeterminacy, until an experiment is carried out to discover it. The answer obtained from such an experiment is then subject to the same regularity of probabilities. Satisfyingly, experiments to determine the quark–antiquark content of π^0, by a study of its interactions, give the answer "u$\bar{\text{u}}$" in 50% of the results, and the alternative answer "d$\bar{\text{d}}$" in the other 50%.

The outcome from a high-energy collision between (say) two protons is subject to quantum indeterminacy in several ways. Imagine that two protons collide with a total kinetic energy of 300 MeV. This is enough energy to make (up to) two pions (remember, they have a mass energy of 140 MeV each) and still have a little energy left over to provide kinetic energy of the products. But what exactly will the products be? In fact there are several possibilities in this case:

$$p + p \rightarrow p + p + \pi^0$$

$$p + p \rightarrow p + n + \pi^+$$

$$p + p \rightarrow p + p + \pi^0 + \pi^0$$

$$p + p \rightarrow p + p + \pi^+ + \pi^-$$

$$p + p \rightarrow p + n + \pi^+ + \pi^0$$

$$p + p \rightarrow n + n + \pi^+ + \pi^+$$

Each of these possibilities will occur, and there is no way of predicting which one will be the outcome of a particular reaction. The rules that must be obeyed in these reactions are:

- Energy is conserved (as usual).
- Electric charge is conserved (as usual).
- The number of quarks minus the number of antiquarks is conserved.

Considering each of these rules in turn for the six reactions above, first look at the conservation of energy. In each case, the mass energy of the two protons on the left-hand side (2×940 MeV) is balanced by the mass energy of the two protons or neutrons on the right-hand side (2×940 MeV). So, the only concern is what happens to the 300 MeV of kinetic energy of the reactants. The mass energy of each pion is 140 MeV, and clearly there is enough energy in the reaction to create either one or two pions. Energy will be conserved as long as the kinetic energy of the products is 160 MeV in the first two reactions (left over after the creation of one pion with a mass energy of 140 MeV), or 20 MeV in the last four reactions (left over after the creation of two pions with a mass energy of 280 MeV). Three pions cannot be created as that would require at least 3×140 MeV = 420 MeV of kinetic energy from the reactants.

The net electric charge of the reactants is twice that of a proton in each case. The net electric charge of the products in the six cases is also twice that of a proton, so electric charge is conserved in all six cases.

Finally, the reactants are composed of three quarks for each proton and no anti-quarks, so the number of quarks minus the number of antiquarks is six for the reactants. In the products, the number of quarks is three for each proton or neutron and one for each pion, while the number of antiquarks is one for each pion. The number of quarks minus the number of antiquarks in the reactants is therefore $(3 + 3 + 1 - 1) = 6$ in the first two cases and $(3 + 3 + 1 - 1 + 1 - 1) = 6$ in the last four cases. Therefore all three rules are indeed obeyed in all six reactions.

In conclusion, just as electrons, protons, and neutrons simplify the elements to their essentials, so a new layer of apparent complexity can be understood by the quark model. However, there is a crucial difference. If an atom is hit hard enough, electrons come out. If a nucleus is hit hard enough, nucleons come out. But if a nucleon is hit hard, with another nucleon, quarks do not come out. Instead the kinetic energy is transformed into the mass of new hadrons.

7.4   A SUMMARY OF THE UNIVERSE AT SMALL SCALES

You have now completed the first part of the book and have reached the end point of the quest to examine the Universe on its smallest size scales. This introduction to the world of quantum physics has revealed that all material objects are composed of atoms, which in turn consist of negatively charged electrons surrounding a positively charged nucleus. The nuclei of atoms are made up of protons and neutrons, which in turn are formed from combinations of up and down quarks. So to build everything in the world – from the food you had for breakfast and the book in front of you, to every person you have ever met, the planet Earth and all the stars and galaxies in the Universe – requires only *three* types of particle: electrons, up quarks, and down quarks. Virtually imperceptible particles, known as electron neutrinos, complete the picture of the material world.

Understanding of these fundamental building blocks is based on quantum physics. This says that there can be no absolute knowledge of the position of an individual particle, when combined with absolute knowledge of the velocity with which it moves. All that can ever be measured is some kind of probability for finding it in a certain place at a certain time. Systems of particles – such as atoms – do have a well-defined energy, however. Their energy is so well defined, in fact, that the energy of an atom is only allowed to have certain specific values. Between these allowed energy levels, dramatic quantum jumps occur.

It is these jumps that provide the link with the other important component of the world around us: electromagnetic radiation. When an atom jumps from one energy level to another, such that it loses energy, a photon is emitted whose energy corresponds exactly to the energy difference between the two energy levels. Alternatively, a photon whose energy corresponds exactly to the energy difference between two energy levels of an atom may be absorbed, so raising the energy of the atom and exciting it to the higher energy level.

Having reached the limits of human understanding at the smallest length scales, it is time to turn to the largest length scales of the Universe and see what surprises lie in store there.

Part II

The Large-Scale Universe

In this second part of the book, some of the key observations in the realm of cosmology are introduced. The structure and contents of our local patch of the Universe are reasonably familiar to many people. The Earth is a rather small, rocky body and is one of several planets that orbit their local star, the Sun. The Sun itself is a fairly average star, about 5 billion years old and midway through its life. It is just one of a hundred billion or so stars that comprise a galaxy known as the Milky Way (or just the Galaxy, with a capital G). The Galaxy takes the form of a flattened spiral structure, with the Sun lying in one of the outer spiral arms, about two-thirds of the way from the centre to the edge. Finally, the Galaxy is merely one of several dozens that comprise the Local Group of galaxies, and the Local Group is itself part of the much larger Virgo Supercluster of galaxies that may contain around 100,000 individual galaxies.

Consideration of the structure and composition of the Universe will be returned to in more detail towards the end of the book, but for now the following three chapters will investigate how the Universe may be observed and two key aspects of the way the Universe appears on its largest scales. The first thing to consider is how we gain information about the Universe through astronomical observations. Following this is an exploration of how the Universe is moving, by looking at the expansion of the Universe. Then, the radiation content of the Universe is considered, with an examination of how the Universe is gradually cooling. As a result, by the end of this second part of the book, you will have explored answers to the second part of the first question posed in Chapter 1 – how does the Universe behave on large scales?

8 Observing the Universe

Apart from the odd meteorite that lands on Earth, or the occasional Moon rock brought back by astronauts, virtually all of the information known about the Universe at large comes from the light or other electromagnetic radiation that may be detected with ground-based or space-based telescopes and detectors. (Although gravitational radiation may now also be detected from some extreme events, as mentioned at the end of this chapter.) Here, we first look at how telescopes operate in the visible part of the spectrum before moving on to consider the different techniques that are used at the long-wavelength and high-energy ends of the electromagnetic spectrum respectively.

8.1 TELESCOPES

Unaided human eyes are not very suitable for astronomical observation. First, the eye has a limited sensitivity – a distant source of light, such as a star, will not be seen at all unless the intensity of light from it reaching your eye is above the sensitivity threshold of the retina. Second, the acuity of the eye is limited by the physical size of the detectors on the retina and by the small aperture of the eye. This limited resolution makes it impossible for human eyes to separate individual distant sources of light that are too close together or to make out details of their shape or structure.

The minimum separation at which objects may be distinguished is referred to as the limit of angular resolution. As you know, there are 360 degrees in a complete circle. Each degree may be subdivided into 60 arcminutes, and each arcminute further subdivided into 60 arcseconds. The typical angular resolution of the unaided human eye is about 1 arcminute and corresponds to (say) the size of a tennis ball seen from over 200 metres away. The angular diameter of the Sun or full Moon seen from Earth are each about 30 arcminutes.

The invention of the telescope at the beginning of the 17th century was an important milestone in the advancement of astronomy. It provided a simple instrument that overcame, at least in part, these shortcomings of human eyes. This section will first consider the various designs of refracting telescopes and reflecting telescopes that have been developed over the past four centuries, and then look at the main ways of characterizing the performance of an astronomical telescope. An important piece of nomenclature is as follows: the name refracting telescope (or refractor, for short) is used to indicate a telescope in which only lenses are used to form the image; the name reflecting telescope (or reflector for short) is used to indicate a telescope in which a curved mirror is used in place of the objective lens.

Although, in this section, telescopes that operate in the visible part of the electromagnetic spectrum (i.e. wavelengths of a few hundred nanometres) will be described, the principles may be applied similarly to telescopes that operate in the near-infrared and near-ultraviolet (i.e. wavelengths of a few micrometres and around 100 nanometres, respectively). These regions of the electromagnetic spectrum are invisible to human eyes, but the radiation can be focused by telescopes and detected electronically in much the same way as described in this section.

8.1.1 REFRACTING TELESCOPES

The story began in 1608 when a Dutch optician called Hans Lippershey discovered (probably quite accidentally) that a distant object appeared larger when viewed through a combination of two lenses: a relatively weak (long focal length) converging lens facing the object and a strong (short focal length) diverging lens in front of the eye. This combination of lenses was subsequently used by Galileo Galilei for looking at the Moon, the planets, and the stars, and it became known as the Galilean telescope. Using his telescope, Galileo was the first to observe that Venus displayed phases, like the Moon, and also the first to see the four large satellites of Jupiter, now referred to as the Galilean moons: Io, Europa, Ganymede, and Callisto.

By about 1630 Johannes Kepler had replaced the diverging eyepiece lens with a converging one of very short focal length. This new combination of two converging lenses is now known as the Keplerian telescope. It has remained the main form of construction for refracting astronomical telescopes until the present day, although many technological improvements have been introduced to cope with the various problems that set limits to the basic telescope's performance.

In order to optimize the performance of an optical telescope, both the aperture and the focal length of its objective lens must be as large as possible. Unfortunately, this is easier said than done. To begin with, there are serious technological problems in producing very large lenses. To ensure that the initial block of glass, from which the lens is to be made, is perfectly transparent and optically homogeneous throughout, the molten glass may need several years of gradual and controlled cooling. Next comes the problem of grinding and polishing – it is not easy to sustain a perfect spherical curvature for a very large focal length lens over the whole of its surface area. And when you have a large lens, it is inevitably a thick lens, which therefore absorbs strongly at the blue end of the spectrum. It is also a very heavy lens, which means that it would have a tendency to sag under its own weight. In practice, usable objective lenses with a diameter much larger than 1 metre cannot be made.

As for the maximum possible value of the focal length, the limits are set not just by the problems of grinding the right curvature, but also by the need to make the whole instrument movable. Clearly the physical length of a Keplerian refracting telescope cannot be less than the focal length of its objective lens, so it would hardly be realistic to plan a telescope with a focal length of 100 metres.

Finally, there is the problem of optical aberrations. Light passing through different parts of a lens is not focused to the same point by spherical surfaces (this is known as spherical aberration). Even from the same part of the lens, waves of different frequency are focused to different points (this is known as chromatic aberration). By combining several lenses of different optical strengths and different refractive indexes these aberrations may be reduced, but the problems are formidable and increase with the increasing size of the lenses and with the angle of the rays with respect to the optical axis. As a result, even the best refracting telescopes have only a relatively narrow field of view within which the resolution is good. The net result of all these problems is that refracting telescopes are no longer built for serious astronomical work.

8.1.2 REFLECTING TELESCOPES

A lens is not the only object that can collect and focus light and so produce visual images. People have known about and used looking glasses (mirrors) for much of recorded history, but it took no less a genius than Isaac Newton to realize how a curved mirror could be used to construct an optical telescope, and that this would overcome some of the most important shortcomings of refracting telescopes. The key is that a concave spherical mirror will reflect parallel rays approaching along its axis of symmetry so that they come together at one point (the focus) lying between the reflecting surface and its centre of curvature.

The main advantage of focusing by reflection is that the angle of reflection is the same for all frequencies in the incident radiation. There is no analogy to the chromatic aberration that takes place in lenses. So, if the objective lens of a telescope is replaced by a reflecting spherical mirror, the chromatic aberration on the input side of the telescope will be automatically and completely eliminated (it still exists in the eyepiece). Also, there is still spherical aberration because rays reflected from the points further away from the axis of symmetry will be focused nearer to the reflecting surface.

Unfortunately, by reducing the size of the mirror to reduce spherical aberration, some of the potential light-gathering power is lost, and the useful field of view is also limited. There are two ways of dealing with this problem. Either a paraboloidal shape may be chosen for the mirror or the focusing of a spherical mirror may be corrected by introducing a suitable pre-distortion into the incoming wave front. This is done by placing in front of the mirror a transparent plate whose shape is such that it refracts the initial parallel rays near the optical axis differently from those further away from it. The correcting plate is known as a Schmidt plate, and the reflecting telescopes in which such a plate is used are called Schmidt telescopes.

In case you are wondering how you could actually see the image of a star produced by a spherical converging mirror without being in the way of the oncoming light, this problem was solved simply and neatly by Newton. He put a small flat mirror (the secondary mirror) just before the focus of the main reflector (the primary mirror) and at an angle of 45° to the optical axis. This moved the image towards the side wall of the telescope tube, where he then fixed an eyepiece for direct observations. A telescope using this arrangement is known as a Newtonian telescope. A further improvement was introduced by the French astronomer Guillaume Cassegrain, who was one of Newton's contemporaries. His idea is now used in many large modern telescopes. In place of Newton's flat and tilted secondary mirror, Cassegrain used a slightly diverging secondary mirror placed on the optical axis of the main one. The light is therefore reflected back towards the centre of the primary mirror, where it passes through a hole on the optical axis and then onto an eyepiece. This has the effect of extending the path of the reflected light before it is brought to a focus. As a result, the effective focal length of the objective mirror (which determines the size of the image) in a Cassegrainian telescope can be roughly double that for a Newtonian telescope of the same length. Both Newtonian and Cassegrainian telescopes may be constructed using either paraboloidal objective mirrors or using spherical objective mirrors with Schmidt correcting plates.

If a telescope is to be used with an electronic detector (see the next section), instead of the eye, then the image must fall onto the light-sensitive surface of the detector. In this case there is no point in using the telescope with an eyepiece lens. The simplest solution is to remove the eyepiece lens entirely and place the detector in the focal plane of the objective mirror. This also has the advantage of removing any aberrations introduced by the eyepiece lens. Alternatively, the secondary mirror may also be removed, and the detector may be placed directly at the prime focus of the main mirror. This has the additional advantage of removing one more optical component, and with it the inherent aberrations and absorption losses that it contributes.

In comparison with refracting telescopes, the reflectors start with the important advantage of zero chromatic aberration. But they also score heavily on some aspects of practical construction and technology. It is much easier to produce mirrors (rather than lenses) of very large diameter (5 m or so) because the glass does not have to be perfectly transparent or optically homogeneous. Arrays of mirrors can also be combined to make even larger collecting areas. Also, the grinding and polishing of a mirror is carried out on only one surface, which is finally covered by a thin reflecting layer of aluminium. On the debit side, there is greater loss of optical intensity than in the refractors, because the reflecting surfaces are never quite perfect and may have appreciable absorption. Aluminized surfaces also deteriorate rather quickly and must be renewed relatively frequently. On the other hand, a perfectly polished lens remains serviceable for many years.

8.1.3 THE CHARACTERISTICS OF ASTRONOMICAL TELESCOPES

Having looked in detail at the different designs of optical telescopes and the various problems inherent in their construction, it is now time to turn to the ways in which their performance may be characterized. There are three main performance characteristics, each of which may be applied to both refracting telescopes and reflecting telescopes.

First, the light-gathering power of a telescope is simply defined as the area of the objective lens (or mirror) divided by the area of the eye's pupil. It quantifies the gain in the amount of light provided by the telescope over the naked eye: bigger telescopes can gather more light and so see fainter objects.

Second, the angular magnification of an astronomical telescope is defined as the focal length of the objective lens (or mirror) divided by the focal length of the eyepiece lens. Telescopes with greater angular magnification can form larger images and so allow more detail to be seen in extended objects.

Finally, the limit of angular resolution of a telescope may also be defined. This is the minimum angular separation of two equally bright stars, which would be just resolved by an astronomical telescope (assuming aberration-free lenses and mirrors and perfect viewing conditions). The limit of angular resolution measured in radians is given roughly by the wavelength of the light divided by the diameter of the telescope's objective lens or mirror. Clearly, larger telescopes can resolve finer detail than smaller telescopes. An optical telescope with a primary lens or mirror that is just 12 cm in diameter would have a theoretical limit of angular resolution of about 1 arcsecond.

8.2 ASTRONOMICAL DETECTORS

Human eyes respond only to the average rate at which light from the source is reaching the retina. Once this rate falls to the threshold of sensitivity, the visibility of such a weak source cannot be improved by gazing at it for a long time. In fact, because of stress and tiredness, the very opposite tends to happen. So, although a telescope with a great light-gathering power undoubtedly helps to discover new faint sources, the sensitivity threshold of the eye is still a limiting factor.

This limit was overcome in the early 20th century by replacing the eye with a photographic emulsion deposited onto a glass plate or film base. By using long exposure times, it became possible to detect sources that are several orders of magnitude fainter than can be detected with the eye. Photographic emulsion is known as an integrating detector because it can add up the light it receives over a long period of time. The only limitation here is the background brightness of the sky itself. If the exposure is too long, the photograph will eventually record the minute intensity of the scattered light in the atmosphere, and the faintest light sources will be lost in this background. In practice, ground-based telescopes using a photographic emulsion as the detector can record stars that are up to 10–20 million times fainter than those the unaided human eye can detect. Although such "wet chemistry" photography was the mainstay of astronomy for over a century, in recent decades electronic imaging detectors have largely replaced it as the means to record astronomical images.

8.2.1 PHOTOMETRY

Photometry means simply measuring the brightness of astronomical objects, and electronic means of doing this are now widely used in astronomy. A photoelectric detector is essentially a device that responds to incoming photons of light by producing an electrical signal. This electrical signal is then detected, amplified, and measured, and the resulting "image" is built up and processed using a computer. Several forms of photoelectric detector have been used over the years, such as *photomultipliers* and *photodiodes*, but nowadays, the commonest type of detector used in astronomy is known as a CCD, which stands for charge coupled device.

A CCD is a two-dimensional, highly sensitive solid-state detector that can be used to generate, extremely rapidly, a pictorial representation of an area of the sky. Physically CCDs are very small, typically only a couple of centimetres across. They are made from a silicon-based semi-conductor, arranged as a two-dimensional array of light-sensitive elements, known as pixels. A modern CCD may contain 4096 × 4096 of these light-sensitive elements (i.e. around 16 million pixels) in a single detector array, each of which is typically only 20 μm across.

The individual elements on the CCD can each be considered as tiny detectors in their own right. When light falls on a pixel, electrons in the semiconductor are released. The number of electrons released depends on the intensity of the radiation, and the accumulated charges are transferred out of the array in a controlled manner, one row at a time. This forms a digital video signal that can be displayed on a screen or stored on a computer for later processing and analysis.

CCDs have an extremely high photon efficiency; they can detect *individual photons* and they record about 60% of the photons that fall on them. This may be compared with the efficiency of photographic emulsion which is typically only a few percent, or the human eye for which it is around 1%.

The disadvantage of CCDs lies in the fact that they are relatively small and so images obtained with individual detectors only cover an area of sky that is typically a few arcminutes across, if they are to have good spatial resolution. Large arrays of CCDs are needed to span greater areas with high resolution, but this is increasingly becoming feasible as the cost of the detectors reduces. For instance, the Legacy Survey of Space and Time (LSST), to be carried out using the 8.4-metre Simonyi Survey Telescope at the Vera C. Rubin Observatory, which will begin operations in 2024, will have 200 CCDs giving a 3.2 billion-pixel image spanning a 3.5 × 3.5 degree area of sky (about 50 times the area of the full moon). It will image the entire visible sky every three nights through six filters, generating 20 TB of data every night for at least ten years.

8.2.2 SPECTROSCOPY

So far, light-sensitive devices placed directly in the focal plane of the telescope have been considered. Used in this way, they act as integrating detectors, providing information about the brightness of stars and other objects, but they do not provide information about the spectral composition of the recorded radiation. All that may be learnt is how different celestial objects compare in overall brightness, over the total range of spectral sensitivity of the detector.

The apparent colours of celestial bodies, or, more accurately, their complete emission or absorption spectra, are determined by their composition, by their temperature, and by the physical processes taking place in them. If this vital information is to be obtained it is necessary to use techniques that separate the radiation according to frequency (or wavelength), before its intensity is recorded or measured. A very simple way of doing this is to use optical filters that absorb certain parts of the spectrum and allow through only a narrow band of wavelengths. In astronomy, broad-band filters are commonly used, labelled as U, B, V, R, and I, which span wavelengths in the near-ultraviolet, blue, green-yellow (visible), red, and near-infrared parts of the spectrum respectively.

A more sophisticated and accurate method for obtaining full spectrophotometric information (intensity as a function of wavelength) about a celestial source of radiation is to decompose its spectrum using either a prism or a diffraction grating. As noted earlier, a diffraction grating is simply a piece of glass with a set of very finely ruled lines etched onto it. A grating used for spectroscopy will typically have several hundred lines per millimetre, so the line spacing is comparable to the wavelength of light. When visible light passes through a diffraction grating, different wavelengths will emerge at specific angles relative to the original direction of travel of the light. As a result, the component wavelengths are dispersed or separated spatially. A given grating will produce a diffraction pattern consisting of multiple orders, each of which contains a spread of all the wavelengths present in the incident light. In the first order of the diffraction pattern, the sine of the angle of diffraction is given by the wavelength divided by the line spacing of the grating.

In a spectrometer, the primary image from the telescope falls onto a screen containing a single (vertical) slit. This ensures that the light from only a vertical strip of the sky, containing perhaps one or a few stars, passes through. The light is then collimated by a lens such that a parallel beam of light then falls on the diffraction grating. This (vertical-line) grating disperses the light horizontally: red light (long wavelength) is diffracted through a larger angle than blue light (short wavelength) as indicated above. Another converging lens then focuses the diffraction pattern onto the CCD detector positioned in the lens's back focal plane.

The CCD is usually positioned to record only the first order of the diffraction pattern. With a monochromatic, point light source, this would correspond to a single diffraction spot. Since the light from a star usually contains a continuous distribution of wavelengths (that is it has a continuum spectrum), the first diffracted order from a star is smeared out into a line, perpendicular to the lines of the grating (horizontally on the CCD).

It should now be clear why a vertical slit is used between the telescope and the grating: only stars lying within the slit can contribute to the diffraction pattern formed on the CCD. If a slit were not used, the spectra of many stars, spaced horizontally in the image plane of the telescope, may lie on top of each other, making interpretation difficult. With the slit perpendicular to the dispersion direction, stars spaced vertically in the image plane of the telescope (in the slit) will produce spectra that do not overlap on the CCD. The appearance of the diffraction pattern on the CCD will therefore be a series of horizontal spectra lying vertically above one another: one spectrum corresponds to each star imaged onto the slit.

What will the spectrum of an astronomical object look like? Well, if the light from the star or other object has only a continuum spectrum, of uniform intensity, the CCD will register a constant intensity across the field (assuming that it responds equally to all wavelengths). If, as is more likely, the star's continuum spectrum varies in intensity as a function of wavelength, then this will appear as a varying intensity across the CCD. The presence of any emission or absorption lines in the spectrum will show up as either bright or dark lines superimposed on the smoothly varying background. By calibrating the whole system, using a spectrum from an arc lamp that contains spectral lines of known wavelength, the wavelength of any emission or absorption line in the spectrum may be determined from its position on the CCD.

The gratings used in spectrometers on astronomical telescopes typically have between 100 and 2000 lines per mm. Gratings with a large number of lines per millimetre disperse the wavelengths over a large distance on the CCD, but consequently only a limited wavelength range can be recorded within the space available. Such spectra are useful for looking at the detailed structure of individual spectral lines and are said to have a high spectral resolution. Gratings with a small number of lines per millimetre, on the other hand, do not disperse the wavelengths by as much on the CCD. However, in this case a much larger wavelength range can be recorded. Spectra of this type are useful for looking at the overall spectral distribution and the shape of the continuum spectrum; they are said to have low spectral resolution. Different regions of the spectrum can be selected to fall on the centre of the CCD, in this way it is possible to select either individual spectral lines or particular regions of the spectrum that are of interest.

The CCD output from a grating spectrometer, or a plot of intensity against wavelength, may not look much like an "image" of the star under observation. Yet, it contains far more information than any "likeness" photograph ever can. To begin with, the wavelengths of the lines, their grouping into series, and their relative intensities identify the chemical elements emitting or absorbing the light. Second, the width of individual lines and their detailed profiles carry information about the physical conditions in which this emission or absorption took place (e.g. the density and the temperature of the star's atmosphere, the turbulences in it, the presence of electric and magnetic fields, and the speed of rotation of the star). And last but not least, the red shift (or blue shift) of spectral lines gives information about the speed at which stars or galaxies are receding from (or approaching towards) the observer, as you will see in the next chapter.

8.3   BEYOND THE VISIBLE

Visible light is only a very narrow band (from about 400 nm to 700 nm) in the spectrum of electromagnetic radiation; it is distinguished by nothing more than the fact that it creates responses in human visual organs. Electronic detectors used with normal optical telescopes can extend this down to about 150 nm in the ultraviolet and up to about 10 μm in the infrared (atmosphere permitting). However, even from Earth-bound experience it is known that sources of visible light also radiate energy at other frequencies. It would be very strange indeed if stars were to radiate only at the wavelengths visible to humans.

With the development of wireless communication it was only natural therefore to search for radio signals coming from outside the Earth, but until about 1930 all such attempts were doomed to failure because the equipment then available was insufficiently sensitive. In the end, as with so many other important discoveries, the first success came more or less by chance.

8.3.1 Radio Astronomy

In the early 1930s an American physicist, Karl Jansky, was working on the problems of radio communications (in the wavelength band around 1520 m) and discovered two important things:

- There was a general "background noise" of radio emission coming from all directions from the Universe.
- Superimposed over this was a noticeably increased noise whenever he pointed his aerial in certain directions in the skies.

This increase could not be explained away as interference of human origin, and although Jansky published a note on this fact, it did not interest him any further at the time. Nor did it interest anybody else sufficiently even to verify the observation. Then about ten years later a radio amateur, Grote Reber, built a radio receiver (working at about 60 cm wavelength) specifically to monitor radio emissions from outside the Earth. Apart from identifying radio waves from the Sun, he found strong radio signals coming from the directions of three major constellations, Sagittarius (the direction from which Jansky had observed increased noise), Cassiopeia, and Cygnus. But even his work did not create much interest and enthusiasm.

These observations were, of course, unwittingly rediscovered many times – not least whenever radio emissions from the Universe affected the operation of radar installations. At the time of the Second World War this effect was just a nuisance to be eliminated, or at worst to be put up with, rather than an interesting problem for serious investigation. It was only after the end of the war that astronomers, physicists, and radio engineers combined their efforts in this field.

By about 1950 radio telescopes were developed to such a degree that it became possible to identify the location of cosmic radio sources rather more accurately than simply to within a particular constellation. Subsequent development was very fast, and today the birth of radio astronomy can be seen as a milestone in the development of astronomy, astrophysics, and cosmology of similar importance to the invention of the optical telescope four centuries earlier. The present state of information about cosmic radio sources can be summarized as follows.

- Most of the strongest radio sources seem to be located in, and can be identified with, optically bright elliptical galaxies. These are usually now called radio galaxies. The two brightest are known as Cygnus A and Centaurus A; the first of these was one of the strong sources noted by Reber. Radio galaxies often exhibit extensive radio jets extending for vast distances into space, beyond the galaxy itself.
- There is a special class of objects, called quasars, which show some peculiar features in their radio and/or their optical emissions. Quasars are thought to be extremely distant galaxies seen at an early stage of their evolution. They contain supermassive accreting black holes in their centres.

- There is a strong radio source, known as Sagittarius A, located at the centre of the Galaxy. (This was the original radio source detected by Jansky and later by Reber.) However, the apparent strength of this source is merely a consequence of its proximity. In general, normal galaxies (such as the Milky Way, the Andromeda galaxy, and others) are only weak radio sources. Indeed, one of the definitions of a normal galaxy is that it shows only weak radio emission; the detection of strong radio emission indicates that something violent is happening within it. Sagittarius A is now recognized to be a black hole with a mass of almost 4 million times that of the Sun, around which many stars are seen to be orbiting under the influence of its strong gravitational field.
- Most of the remainder of the strong, localized, radio sources are supernova remnants within the Galaxy. The strongest is Cassiopeia A (discovered by Reber), and the second strongest is the Crab Nebula (known to radio astronomers as Taurus A). These are the clouds of debris left behind when massive stars explode and which persist for a few thousand years.
- One of the commonest classes of radio source in the Galaxy is radio pulsars, which are recognized as rapidly rotating neutron stars. The Crab Pulsar sits in the middle of the Crab Nebula, but many other pulsars outlive their supernova remnants which have long since dispersed.
- Stars are generally weak sources of radio emission. The Sun (a fairly normal star) appears as a bright radio source only because it is extremely close.

One aspect common to all strong radio sources is the evidence of violent changes having taken place, sometimes following earlier large-scale explosions. These various types of object will be explored further in Chapter 17.

8.3.2 Radio Telescopes

The most serious problem the pioneers of radio astronomy had to face was the accurate location of cosmic radio sources. This difficulty is a result of the inherent properties of radio waves, in particular their wavelength. Radio waves are electromagnetic radiation, as is light, so it would seem reasonable to expect that they should behave in a similar way when reflected. In principle at least, it should be possible to focus radio waves by reflection using spherical or parabolic mirrors, just as light is focused in optical reflectors. But, by the same argument, it must also be expected that if such an instrument is built, its limit of angular resolution would be determined by the ratio of the wavelength to the telescope diameter, just as it is for light.

Since radio wavelengths are of order a billion times larger than optical wavelengths, a radio telescope based on the same principle as an optical reflector appears to be a hopeless instrument as far as angular resolution is concerned. You will see later how this limitation can be overcome. However, radio telescopes do have some important advantages over their optical equivalents. One advantage is that radio emission from the Universe can be studied continuously, day and night. It is not swamped by the Sun's radiation during the day and it penetrates even through a cloudy atmosphere. The second advantage is that the reflecting dish for radio waves does not have to be of the same quality as the mirror in an optical telescope. As a rule of thumb, you could say that imperfections and holes that are significantly smaller than the wavelength to be reflected do not matter. Thus, the dishes of radio telescopes, particularly for wavelengths of several metres and more, do not even have to have continuous surfaces. In practice, a diameter of about 80–100 m represents a reasonable limit of size of a single-dish radio telescope, if it is to remain manoeuvrable. The largest single-dish telescope ever built has a diameter of 300 m and was constructed by making use of a natural bowl of suitable shape that exists in the landscape near Arecibo in Puerto Rico. The Arecibo radio telescope obviously cannot be steered, but the angle of view can be varied, within relatively narrow limits, by changing the position of its aerial within the focal region.

What sort of an image does such a radio telescope produce? In theory, if you consider a large, extended source with distinct spatial variations of luminosity in the radio wave region of the spectrum, the radio telescope should produce in its focal plane a radio image that will show variations in intensity corresponding to the distribution of luminosity in the object. But how can this image be "seen"? Well, in theory again, a point-size detector could be moved around within the image field, and the intensity of radio waves at each point in the radio image could be measured. In this way a numerical reading on some arbitrary scale could be allocated to each point. Finally, this numerical scale could be translated into a scale of optical contrast and so a visual representation of the radio image could be produced. The problem is that even a dish that is 100 m in diameter can only resolve angular details larger than about 40 arcminutes at a wavelength of 1 m (or larger than 4 arcminutes at a wavelength of 10 cm). But very few celestial objects have angular sizes larger than a few arcminutes; most are only measured in arcseconds.

So, in practice, this is *not* the way that radio images are obtained. Instead, a more complex technique is used, as described in the next section. Single-dish radio telescopes are generally only used to provide information about the total intensity or polarization of a source of radio waves at different wavelengths, rather than a map of its spatial structure.

8.3.3 Radio Interferometry

As discussed above, it is not possible in general to obtain resolved radio images of individual radio sources from single observations by single radio telescopes. However, such images *can* be built up gradually by making use of radio interferometry and aperture synthesis. Radio interferometers use (at least) two separated radio dishes to collect the radiation. The two electrical signals, collected by small aerials at the focal points of the telescopes, are fed through identical cables to a common receiver. The combined output of this receiver will depend on the aiming angle and on the separation between the dishes (known as the baseline), which can be varied by moving one of the telescopes. Real celestial objects appear as two-dimensional projections on the sky and good two-dimensional radio images may be obtained by varying the *orientation* of the base of the interferometer, as well as the length of the baseline. Each of the orientations gives information about the brightness of the source along that particular direction, and by combining them all, the whole image may be built up. This is known as aperture synthesis, because the net effect is to synthesize a telescope aperture that is the same size as the distance between the two individual radio dishes. Angular resolutions of far better than one arcsecond can be achieved in complete, aperture-synthesized, two-dimensional radio images.

There are two basic approaches to aperture synthesis. The first method is to have a large number of identical interferometers (pairs of radio telescopes) working simultaneously at different (fixed) orientations and different (fixed) baselines. The second method is to have just one pair, which goes through different orientations and baselines sequentially. The advantage of the first approach is that it is much faster, but it is, of course, more expensive, and therefore in practice limited to relatively small telescopes working with relatively short baselines. The second approach is technically simpler, particularly if the natural rotation of the Earth is used to do the job of turning the base around, but it cannot detect any short-time variations in the distribution of brightness.

In practice, most linked arrays of radio telescopes, such as the Multi-Element Radio Linked Interferometer Network (MERLIN) in England, the Low Frequency Radio Array (LOFAR) centred on the Netherlands and stretching across Europe, or the Very Large Array (VLA) in New Mexico, combine both of these techniques (Figure 8.1). During the course of a single observation, each of these arrays uses many different telescopes with different, fixed, baselines and fixed relative orientations *and* they make use of the rotation of the Earth to provide a whole range of orientations with respect to the radio source.

With the VLA, the 27 identical radio telescopes are mounted alongside railway tracks in a Y-formation. The dishes may be moved along the tracks to a series of different locations, thus giving a range of baselines that may be used in different observations. This in turn allows astronomers to obtain different spatial resolutions and fields-of-view. Each dish is 25 m across, and the maximum effective aperture of the array (from the end of one leg of the Y to the end of any other leg) is about 34 km. Every eight hours, the rotation of the Earth means that each arm of the Y-shaped array lies in the direction previously occupied by one of the other arms. Hence the complete aperture of the telescope array is synthesized in an eight-hour observation period. When the 27 telescopes are *closely packed*, images with low angular

FIGURE 8.1   The VLA radio telescope array in New Mexico, USA.

resolution but covering a large field-of-view are obtained. Conversely, when the 27 telescopes are *widely spaced* along the arms, this results in images with high angular resolution but only a small field-of-view.

For special purposes, when full synthesis is not required, it is possible to set up very long baseline radio interferometers, by linking together pairs of radio telescopes separated by intercontinental distances. This technique is known as very long baseline interferometry (VLBI). The outputs of the two or more telescopes are recorded independently, with an agreed set of time marks for later synchronization. The recordings are then fed into a computer programmed to perform their interferometric superposition and analysis. Such long baseline interferometers have achieved angular resolutions as small as half a milliarcsecond, so beating even the *theoretical* resolution of the largest optical telescope by two orders of magnitude, and the *practical* resolution by three orders of magnitude. (Remember, a milliarcsecond is 1/1000 of an arcsecond, and there are 3600 arcseconds in 1 degree.)

8.4 ASTRONOMY AT HIGH ENERGIES

In "laboratory" physics there is general agreement that X-rays and gamma-rays should be distinguished strictly according to their origin. X-rays come from transitions of electrons between the discrete energy levels they are allowed to occupy within the structure of atoms, whereas gamma-rays originate inside the nucleus as a result of nuclear transitions. However, in many branches of applied physics and technology, a rather more pragmatic distinction is usually applied: X-rays are arbitrarily defined as photons with wavelengths between (very roughly) 3×10^{-8} m and 3×10^{-12} m, and gamma-rays are all photons of wavelength shorter than about 3×10^{-12} m. This distinction according to wavelength, rather than according to origin, is also customary in astronomy – mainly because there is usually no way of knowing about the origin of the radiation anyway.

As you have seen, another custom is to talk about the energy of X-ray and gamma-ray photons, rather than their wavelength or frequency, and to express the energy of these photons in electron volts. For instance, a photon energy of 1000 eV or 1 keV is equivalent to a frequency of approximately 2.4×10^{17} Hz or a wavelength of about 1.2×10^{-9} m. The dividing line between the energies of X-rays and gamma-rays, corresponding to the wavelength given in the previous paragraph, is of order a few hundred kiloelectronvolts.

8.4.1 X-Ray and Gamma-Ray Astronomy

Cosmic X-ray detectors must be operated outside the Earth's atmosphere as X-rays from outer space are absorbed at high altitudes. As a result, high-energy astronomy only began in the 1960s with a series of detectors flown on sounding rockets. These experiments were carried into the upper atmosphere to give just a few minutes of observation, before the capsule carrying the payload returned to Earth hanging from a parachute. A few bright X-ray sources were located in this way, but in general they could not be localized very precisely. In 1970, the first dedicated X-ray astronomy satellite was launched. Known as *Uhuru*, the mission operated for over 2 years and detected over 300 cosmic X-ray sources. In the following decades, many other X-ray and gamma-ray astronomy satellites followed, with ever-increasing sensitivity, spatial resolution, and spectral resolution.

Astronomical sources of X-rays and gamma-rays tend to indicate the most extreme environments in the Universe. The main categories of object emitting such radiation can be broadly categorized as follows.

- The most luminous (but distant) X-ray and gamma-ray sources are the nuclei of active galaxies (active galactic nuclei or AGN for short), in which supermassive black holes accrete vast quantities of material from their surrounding environments.
- Within distant clusters of galaxies there are often pools of hot gas falling into the centre of the cluster, and these too glow with X-ray intensity.
- Closer to home, within the Galaxy, the strongest X-ray and gamma-ray sources are compact binary stars in which a dead stellar remnant (a white dwarf, neutron star, or black hole) is pulling material off its companion star and accreting the transferred matter.
- The clouds of gas blown off by supernovae (supernova remnants mentioned earlier) are hot enough that they glow strongly in X-rays for a few thousand years.
- Normal stars are generally weak X-ray sources, unless they are rapidly rotating, in which case, coronal X-ray emission from their outer regions may become strong enough to observe.
- The Sun is intrinsically a weak source of X-rays and gamma-rays but appears bright at these energies merely because it is so close.

There will be more to say about these classes of object in Chapter 17.

8.4.2   X-RAY AND GAMMA-RAY TELESCOPES

The detection of this extremely short-wave (high-energy) radiation requires completely different techniques from those described earlier. The glass of optical telescopes, which is almost perfectly transparent within the visible range of wavelengths, becomes almost completely opaque to X-rays and gamma-rays. Similarly, metallic surfaces (reflecting telescopes or radio telescopes) are much more likely to absorb than to reflect these high-energy photons, so it is not possible to use the same sort of telescopes as are used in the optical range.

Early X-ray and gamma-ray telescopes were little more than "photon buckets," collecting all the high-energy photons which arrived at the detector from one (rather large) patch of sky. Their angular resolution was extremely poor – often of the order of a couple of degrees or more. Since the early 1980s however, two important techniques have been developed which overcome this limitation to some extent.

X-rays will be reflected from the inner surfaces of highly polished, gold-coated cylinders if the X-rays strike the cylinders at very shallow, grazing incidence, angles. By tapering these cylinders into paraboloidal and hyperboloidal cross-sections, the X-rays can be focused onto position-sensitive electronic detectors at the end of the "cylinder telescope." The target area presented to the sky by just one of these cylinders is rather small, and a larger effective area is produced by stacking many of these X-ray mirrors inside one another. All the concentric mirrors focus X-rays onto a single X-ray detector, placed in the focal plane of the telescope. Practical X-ray telescopes can produce X-ray images that are of the order of a degree across, with an angular resolution of only a few arcseconds.

At the time of writing, NASA's *Chandra* X-ray observatory and ESA's XMM *Newton* observatory are the workhorses of X-ray astronomy (Figure 8.2). Both satellites use grazing incidence optics and have been operating for over 20 years, providing X-ray images and spectra of thousands of cosmic X-ray sources.

Grazing incidence mirrors cannot be used for gamma-ray astronomy as the gamma-rays are simply absorbed by the mirrors. Instead a radically different technique, based on the ideas of a pin-hole camera, is used. The more "pin-holes" you have, the more radiation is passed to the detector and the brighter the image produced. The technique of coded aperture imaging takes this to the extreme. A gamma-ray telescope of this design has a complex coded aperture mask, made from tungsten, which is mounted on the front of the device. The mask is 50% transparent and 50% opaque, with a complex, but regular arrangement of elements. The gamma-ray source effectively casts a shadow of the mask onto the detector. By deconvolving the image with the mask pattern it is possible to deduce a map of the gamma-ray source(s) on the sky that the telescope is looking at. Other gamma-ray telescopes, operating at even higher energies, track the incidence direction of individual gamma-rays directly, from the pair creation events they trigger in electronic detectors, described in the next section.

Two current observatories making use of these techniques are ESA's *International Gamma-Ray Astrophysics Laboratory* (INTEGRAL) and NASA's *Fermi Gamma-ray Space Telescope*. This pair of satellite observatories have also logged more than a decade of continuous operations each.

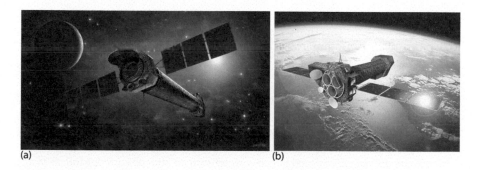

(a)   (b)

FIGURE 8.2   The *Chandra* (a) and *XMM-Newton* (b) X-ray observatories. (Credit: NASA/ CXC and J. Vaughan (a) and ESA/D. Ducros (b).)

8.4.3 HIGH-ENERGY PHOTON DETECTORS

Having focused the X-rays or gamma-rays onto the image plane of a telescope, the next question to address is, how can that image be made visible? Clearly it is not possible to simply put human eyes in position and "see" the image – the X-ray photons would not be registered by the retina. The simplest solution is to use an X-ray-sensitive photographic emulsion, and this is often what is used if you have a simple medical or dental X-ray taken. However, sending photographic plates or films into orbit onboard X-ray or gamma-ray satellites would make it rather difficult to get the information back down to Earth, and a more satisfactory solution is to use some form of electronic detector. The information from the detector is then encoded into a radio signal from the satellite which can then be transmitted to the ground, and this allows the relevant image to be reconstructed back on Earth. In doing so, far more information about the X-ray and gamma-ray photons that are detected may be gained.

The very high energy content of individual X-ray and gamma-ray photons, and their strong interaction with the atoms and molecules of almost any material medium, make it possible to detect the impact of each separate photon and to measure its total energy. Such detection techniques have been developed over the past 50 or so years by nuclear physicists and radiologists, and astronomers have been able to modify these techniques so that they can be used on rockets and satellites. There are three different mechanisms of interaction between a high-energy photon and the atoms or molecules of the medium it strikes:

- The photoelectric effect, in which the photon is absorbed and its total energy is spent on releasing an electron.
- The Compton effect, in which a photon scatters from an electron, exchanging energy with it; only part of the photon's energy is used up in freeing an electron and the rest is re-emitted as a lower-energy photon.
- Pair creation, in which the photon disappears while interacting with nuclear electric fields inside the atom. All its energy is transformed into the creation of an electron/positron pair.

All three processes involve the emergence within the photon detector of one or two electrons with considerable kinetic energy. The energy of these electrons is quickly spent on ionization, that is, by detaching further electrons from their bonds within the atoms or molecules of the detector. As a result, the impact of the photon leads in the end to the production of a number of separate, charged particles (electrons and positively charged ions) within the volume of the detector. These charges can be collected and the corresponding impulse of electric current can be registered by electronic devices. The duration and the shape of these impulses depend on the material of the detector, but for a given detector it is possible to relate the amplitude of the impulse to the energy of the registered photon. As a result, both the position of the photon and its energy (and hence wavelength) can be measured. Modern X-ray and gamma-ray telescopes can therefore be used to obtain images, spectra, and time profiles, just like optical telescopes.

8.5   MULTI-MESSENGER ASTROPHYSICS

The world of astrophysics entered a new era on 17 August 2017. On that day, for the first time, a simultaneous signal was detected from an astronomical object using both electromagnetic radiation *and* gravitational radiation (about which you will read more in Chapter 14). The event in question was the merger of a pair of neutron stars in the distant galaxy called NGC4993. The pair spiralled together, then coalesced in a fraction of a second, converting some of their mass into energy and releasing a chirp of gravitational radiation *and* a flash of gamma-rays. The short gamma-ray burst was subsequently observed as a kilonova over the following days and weeks as the glow of the explosion faded, producing further emission across the electromagnetic spectrum. This discovery marked the start of a new phase in the way we observe the Universe and understand how it works.

9 The Expanding Universe

The deduction that the Universe is expanding is based on measurements of two quantities for each of thousands of galaxies: their distance away and the apparent speed with which they are moving. Each of these quantities is determined in a quite straightforward manner by applying laws of physics that are tried and tested here on Earth – but applying them to situations on a much larger scale of both distance and time. In order to make any sense of the observations that will be discussed, it is necessary to assume that the laws of physics that operate in distant parts of the Universe (distant in both time and space) are the *same* as those that operate in laboratories on the Earth, today. In fact, this is only an extreme version of an assumption that underlies the whole of science: it is assumed that the laws of physics were the same in Birmingham yesterday as they will be in Bangalore tomorrow, for instance. If this were not true, then no further progress would be possible. Conversely, the fact that apparently sensible conclusions can be reached by making just this one assumption tends to indicate that it is not such a bad assumption after all. If such assumptions were to lead to inconsistencies with observations, then we'd have to re-examine the original assumptions and possibly modify the laws of physics as they are currently expressed. This process is the essence of the scientific method.

Virtually everything that is known about the properties and behaviour of the Universe at large has been learnt from the light and other electromagnetic radiation emitted by distant stars and galaxies. When light from a galaxy is collected using telescopes, different types of measurements can be made on it. The simplest measurement is to determine how bright the galaxy appears to be, that is how much light emitted by the galaxy is detected here on Earth. A slightly more complex measurement is to examine the spectrum of light emitted by the galaxy. As you will see in the rest of this chapter, it is measurements of the brightness and the spectrum of a distant galaxy that can lead to determinations of its distance and apparent speed, respectively.

9.1 THE DISTANCES TO GALAXIES

Measuring the distances to galaxies is not straightforward however, and in fact the whole process of determining astronomical distance scales is constructed on a range of techniques, each one of which builds on the previous one, in a kind of cosmic distance ladder.

9.1.1 GEOMETRICAL METHODS

Within the Solar System, distances to the Moon and to other planets can be measured by radar ranging. All we need to do is to bounce a radio signal off the distant object and measure how long the signal takes to come back. Then the distance away may be calculated as half the time taken for the round trip multiplied by the speed of light. The speed of light is known very accurately and may be expressed as about 3 × 10^5 km s^{-1}. A distance of one light-second is therefore around 3 × 10^5 km, and one light-year corresponds to about ten thousand billion kilometres (10^{13} km).

Using this information, and knowledge of Kepler's laws (which you will meet in Chapter 14), the average distance from the Earth to the Sun can be calculated precisely. This distance is so useful, it's given a special name – the astronomical unit (AU) – which is equal to about 1.5 × 10^8 km. Alternatively the distance from the Earth to the Sun may be expressed as about 8 light-minutes.

Having determined the size of the Earth's orbit around the Sun in this way, geometry may be used to extend the cosmic distance ladder to the rest of the Galaxy using a phenomenon called parallax. You can appreciate parallax for yourself in the following manner. In a location where you can see to quite a distance away, hold up one hand at arm's length with the thumb extended vertically upwards. Look at your thumb with only your left eye, keeping your right eye closed. Notice where your thumb aligns with distant objects. Then close your left eye, open your right eye, and notice how your thumb appears to jump to the left with respect to the distant objects. That is the effect of parallax – viewing a nearby object from different locations with respect to a distant background.

The same effect can be seen on astronomical scales (Figure 9.1). When nearby stars are viewed from positions in the Earth's orbit that are six months apart, they will be observed to shift by a tiny angle with respect to more distant background stars. Measuring the angular shift, and knowing the size of the Earth's orbit, the distance to the nearby stars may be calculated using simple trigonometry.

If *half* the angular shift over the course of six months is measured to be 1 arcsecond (i.e. 1/60 of an arcminute where 1 arcminute is itself 1/60 of a degree), then the nearby star is said to be at a distance of 1 parsec (short for "parallax arcsecond"). The parsec is therefore equal to: 1 AU/sine (1 arcsec) which is about 3.1 × 10^{13} km or around 3.3 light-years. In fact, no stars are as close as 1 parsec (the nearest star system, Alpha Centauri, is about 1.3 parsecs away), but in general the distance in parsec is equal to the reciprocal of the parallax angle in arcsec. More distant objects have smaller parallax angles, and the most distant objects (beyond our Galaxy) have no measurable parallax at all.

The European Space Agency's *Gaia* satellite launched in 2013 had a mission to map the positions of a billion stars in the Galaxy with unprecedented accuracy. It can measure stellar parallaxes as small as one-tenth of a milliarcsecond (one ten-thousandth of an arcsecond) and so measure the distances of objects as far away as 10,000 parsecs (10 kiloparsecs or 10 kpc), which is about as distant from us as the centre of the Galaxy.

For comparison, the overall diameter of the Milky Way is about 40 kpc and one of its nearest neighbours, the Andromeda galaxy, is about 660 kpc distant. Clusters of galaxies typically lie at distances of several hundred million parsecs away (one million parsecs is written as 1 Mpc or 1 megaparsec), or even several billion parsecs away (one billion parsecs is written as 1 Gpc or 1 gigaparsec).

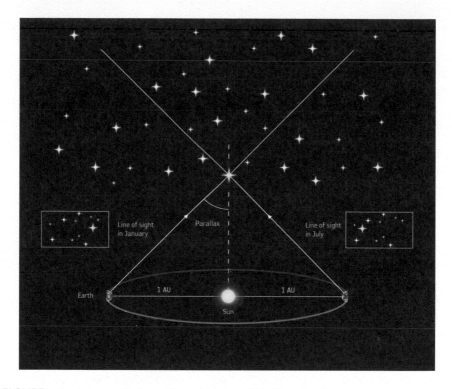

FIGURE 9.1 Astronomical parallax. (Credit: ESA/ATG medialab.)

9.1.2 FLUX AND LUMINOSITY

The parallax technique can only be used within the Milky Way, so how is the cosmic distance ladder extended to the limits of the Universe? As previously mentioned, the only information that may be used is the light from the object and (specifically) how bright it appears to be. In certain circumstances this turns out to be enough information to work out the distance. The reason is that the brightness of an object depends on two things: the amount of light it emits (this is called its luminosity – a measure of the power in watts), and how far away it is. To appreciate this, consider the following thought experiment.

Imagine that you have two identical lamps – both have the same luminosity, say 100 watts, and both emit their light in all directions. You switch them on, then place one of them 100 m away from you, and the other only 20 m away. In the dark, how can you tell which of the two lamps is nearer to you? As you probably realize, the nearer lamp will appear to be brighter than the one that is more distant. So, even though the two lamps have the same luminosity, they have different brightnesses because they are at different distances away from you.

To appreciate the relationship between the luminosity and brightness of an object, consider two astronomical objects, which have the same luminosity, but are at different distances from the Earth. As the light from each object travels out into space in all directions, so the light spreads out over the surfaces of imaginary spheres, centred on the object. By the time it reaches the Earth, the light from the more distant object is spread out over a larger sphere than the light from the nearer object. So, the more distant one will appear less bright than the nearby one because its light is spread out over a bigger area.

In fact, it is convenient to *define* the brightness of an astronomical object to be equal to its luminosity divided by the surface area of an imaginary sphere whose radius is equal to the distance from the Earth to the object. The surface area of a sphere depends on the radius of the sphere squared. So, if the radius is doubled, the surface area increases by a factor of four, and if the radius increases by a factor of three, the surface area increases by a factor of nine.

The word flux is often used instead of "brightness," so the flux of light received from an astronomical object is equal to the object's luminosity divided by the area of the sphere over which the light has spread as it travels towards the detector. The standard unit for flux or brightness is therefore watts per square metre (W m^{-2}). This relationship between brightness and luminosity is said to be an inverse square law. Any relationship where one quantity *decreases* as the square of another quantity *increases* can be classified in this way. You will meet two other examples of inverse square laws later in the book.

9.1.3 STANDARD CANDLES

Distances calculated using flux and luminosity measurements rely on astronomical objects called standard candles, that is objects of *known* luminosity. If the brightness is measured, and the luminosity is known, the distance may be calculated.

In the 1890s, Scottish astronomer Williamina Fleming and the American Edward Pickering, working at Harvard Observatory, observed certain types of pulsating stars known as RR Lyrae variable stars, named after the prototype object in the constellation Lyra. These are easily recognizable as they vary in brightness in a characteristic manner, brightening and fading periodically over the course of typically half a day. They noted that *all* the RR Lyrae stars in a given star cluster had the *same* average brightness. Since all the stars in the same star cluster are at approximately the same distance away from us, this indicated that RR Lyrae stars must all have the same average luminosity. Once a few nearby RR Lyrae stars had their distances measured using their parallax, the average luminosity of all RR Lyrae stars could then be calculated (from their known distances and observed brightnesses). More distant ones could then be used as standard candles to determine the distances to more remote star clusters, using their observed brightnesses and known luminosities. RR Lyrae stars are around 50 times more luminous than the Sun and so can be detected out to distances of about 1 Mpc, within the local group of galaxies. They can therefore be used to extend the cosmic distance ladder about one hundred times further away than parallax measurements.

For the next rung in the ladder a different standard candle is used. In 1908 another Scottish astronomer, Henrietta Swan Leavitt, also working at Harvard Observatory, identified a set of Cepheid variable stars in the Magellanic clouds, the nearby satellite galaxies to the Milky Way. Cepheid variables (named after the prototype star Delta Cephei discovered in 1784) are also pulsating stars, with periods of order tens of days, and again are quite easily recognized. What she discovered is that the Cepheid variables with the longest pulsation periods were on average brighter than the ones with shorter periods. Since all the Cepheid variables in each of the Magellanic clouds are at a similar distance away, this relationship between period and brightness is actually a period–luminosity relationship. In this case, by determining the distances to the Magellanic clouds using RR Lyrae stars, the precise luminosity of Cepheid variables as a function of their period could be determined, from their measured brightnesses and known distances. Then, when even more distant Cepheid variables were measured, their observed brightness and predicted luminosity (from the calibrated period–luminosity relationship) could be used to determine their distance away. Cepheid variables have luminosities that are up to 100,000 times larger than that of the Sun and can be observed at distances up to 40 Mpc away, extending the cosmic distance ladder 40 times further again than the most distant RR Lyrae variables. Just why certain stars pulsate will be considered in Chapter 17.

To extend the cosmic distance ladder still further, even more luminous standard candles are needed. A type of exploding star called a type Ia supernova is the solution. It transpires that such objects, about which you will read more towards the end of the book, also have a standardizable luminosity, which is about 5 billion times brighter than the Sun. By observing the rate at which the light from these supernovae fades after the explosion, they can be scaled to the same luminosity, which is itself determined by observing type Ia supernovae in relatively nearby galaxies whose distance has previously been determined from their Cepheid variables. If a type Ia supernova is observed in a more distant galaxy, its distance may be calculated from the observed peak brightness of the supernova and its calculated standard luminosity. Type Ia supernovae are so luminous, they can be observed in galaxies up to 1 Gpc away.

Supernovae are random events however – typically occurring only once per century per galaxy. Consequently, astronomers cannot determine the distance to a remote galaxy just by sitting and watching it, waiting for a type Ia supernova to explode. Instead, one final standard candle may be used, namely the brightness of an entire galaxy itself, as described below.

Clusters of galaxies can contain anything from a few dozen to a few thousand individual galaxies. When cosmologists look at a cluster, they see that all the galaxies within it have different brightnesses. Now, on the scale of the Universe it is usually adequate to assume that all galaxies within any individual cluster are at about the same distance from us. (Remember that the distances between individual galaxies in a cluster – a few hundred kiloparsecs – are small when compared with the distance to the cluster itself – usually hundreds of megaparsecs.) So, the variation in brightness of galaxies *within* a cluster must reflect an *intrinsic* variation in luminosity from one galaxy to the next. The assumption that cosmologists make is that, wherever they find reasonably large clusters (say more than a hundred members), the *tenth brightest* galaxy in any one cluster has roughly the same luminosity as the tenth brightest galaxy in any other cluster. The tenth brightest is therefore assumed to be a typical galaxy for any cluster and is a final standard candle in the cosmic distance ladder. Although details vary depending on the region of the spectrum in which the luminosity is measured, a typical value for the luminosity of the tenth brightest galaxy in a cluster is around 10^{41} W.

The idea that the tenth brightest galaxy in any cluster of galaxies is a standard candle of constant luminosity has been checked by more direct means in nearby clusters where type Ia supernovae have been observed, so allowing the distance to be determined and the galaxy luminosity to be calculated. These measurements indicate that the assumption of the tenth brightest galaxies all having roughly the same luminosity is valid for nearby clusters. Therefore, it can be assumed that the luminosity of the tenth brightest galaxy in *any* cluster is the same, and so the distance to *any* cluster can be found by measuring the brightness of its tenth brightest member.

9.2   THE APPARENT SPEED OF GALAXIES

This section considers how to measure the apparent speed with which a galaxy is moving with respect to us. Once again, this is based on measurements made on the light emitted by a galaxy, but this time what needs to be examined is the spectrum of the light – how it is distributed with wavelength – rather than the total amount of light emitted. It may not be immediately obvious what spectra have to do with speed measurements, but this will soon become apparent.

When the light from stars or galaxies is spread out to form a spectrum, the spectra are seen to contain many dark lines superimposed on the overall bright background. These are absorption lines and are due to the presence of particular types of atoms. As noted earlier, each type of atom absorbs light of particular wavelengths. The bright, continuous background in the spectrum from a star is produced by photons coming from deep in its atmosphere. As these photons emerge through the star's cooler, outermost layers, photons with specific energies are absorbed by atoms. The absorption lines are therefore characteristic of the particular elements that are present in the outermost layers of the star.

When the spectra of distant galaxies are examined, similar absorption line spectra are seen to those of stars in the Milky Way. It is not surprising that such spectra are rather similar, because the spectrum of a galaxy is simply the sum of the spectra of the billions of stars of which it is composed; because most galaxies are far away, telescopes are unable to distinguish individual stars within them. However, the *positions* of the lines in the spectra of distant galaxies, relative to their positions in the star spectrum, are different. In particular, the lines in the spectra of distant galaxies are all displaced to longer wavelengths, relative to those in nearby stars.

Shifted wavelengths have a very natural interpretation in everyday life. The phenomenon is known as the Doppler effect, and it is probably familiar to you in the context of sound waves, although it applies equally to any wave motion, including electromagnetic radiation such as light (Figure 9.2). The Doppler effect with sound is perhaps most noticeable when an approaching ambulance sounds its siren or as a speeding car races past. As the vehicle approaches and then recedes, apart from growing louder and then fainter, the pitch of the sound is perceived as higher when

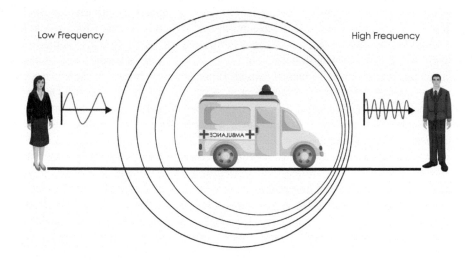

FIGURE 9.2 The Doppler effect with sound waves.

the vehicle is approaching than when it is receding. This can be understood in terms of the sound waves getting "bunched up" in front of the vehicle as it approaches, and "stretched out" behind the vehicle as it recedes. This happens simply because the vehicle moves between the time it emits a particular crest of the sound wave and when it emits the next crest. The bunching up of the waves in front of the vehicle causes the wavelength of the sound reaching your ears to be shorter than if the vehicle were stationary, and the stretching out of the waves behind the vehicle causes the wavelength to be longer.

A shorter wavelength of sound implies a higher pitch of the sound. So, as the vehicle approaches, you hear a higher pitch than if the vehicle were stationary. Conversely, as it recedes from you, the pitch will be lower than if the vehicle were stationary. A similar effect is observed with light. Whereas the wavelength of sound waves may be appreciated by the pitch of the sound perceived by the human ear, the wavelength of a light wave is perceived by the human eye as the colour of the light. A shift in the colour to longer wavelengths (towards the red end of the spectrum) is an indication of motion away from the observer, a shift in the colour to shorter wavelengths (towards the blue) is an indication of motion towards the observer. The lines in the spectra of distant galaxies are shifted towards longer wavelengths, i.e. towards the red.

Astronomers say that such galaxy spectra display a redshift, and an interpretation of this is that distant galaxies are moving away from the Earth. Spectral lines produced by atoms in distant galaxies have the same wavelength as lines produced in similar atomic processes in an Earth-based laboratory. But if a galaxy is moving away from the Earth, the wavelengths of its lines observed on Earth will be shifted towards the red. By exactly the same reasoning, a galaxy spectrum in which features are shifted towards shorter wavelengths, known not surprisingly as a blueshift, would indicate that a galaxy is moving towards the Earth.

The redshift, or blueshift, is defined as the *change* in wavelength divided by the *original* wavelength. The original wavelength is assumed to be that which would be produced by the same type of atoms in a laboratory on Earth. It is also sometimes known as the rest wavelength because it is the wavelength that would be observed from a stationary source. To identify which types of atoms produced the absorption lines in a galaxy spectrum, a certain amount of pattern matching is required to compare whole series of lines rather than just one or two individual lines. The same value of the redshift must apply to *all* lines in a spectrum of a certain galaxy, whatever their individual wavelengths.

The natural interpretation of observing a redshift in the spectrum of a galaxy would be that the galaxy in question is speeding away from us. However, as you will see shortly, this naive interpretation turns out to be not quite the correct picture of what is happening in the Universe. Nonetheless, it is often convenient to convert the measured redshift into an apparent speed of motion. For the speeds that will be considered here, the apparent speed of motion of the galaxy is equal to the redshift multiplied by the speed of light. So, if the wavelength of an absorption line in the spectrum of a galaxy is measured and compared with the wavelength of the same spectral line, as measured in a laboratory, the redshift of the galaxy can be calculated. This can then be converted into an apparent speed of recession (i.e. motion away from us).

As noted earlier, our galaxy, the Milky Way, is one member of a small family of nearby galaxies known as the Local Group. Within the Local Group, a variety of redshifts and blueshifts are observed. This indicates that, in the local neighbourhood, the galaxies are milling around in a fairly random manner. However, if galaxies and clusters of galaxies that are more distant than the Local Group are observed, a remarkable effect is seen: *all* the galaxies exhibit redshifts; *none* exhibit blueshifts. This is remarkable because it shows that *all* clusters of galaxies in the Universe are receding from the Local Group of galaxies! As you will see soon see, this is *not* proof that the Earth lies at the centre of the Universe.

9.3   THE HUBBLE RELATIONSHIP

You've seen that the apparent speeds of distant galaxies can be determined by using redshift measurements, and that their distances may be calculated by comparing the brightnesses of galaxies with their luminosities. When these results are examined for a large number of clusters of galaxies, a quite startling relationship becomes clear: the further away a galaxy is, the larger its apparent speed of motion away from us.

The first person to point this out was the American astronomer Edwin Hubble in 1929. The Hubble relationship may be expressed by the simple statement: the apparent speed with which a galaxy is receding is equal to its distance away multiplied by a number, known as the Hubble constant (Figure 9.3).

Because the apparent speed of a galaxy is typically measured in the unit of km s^{-1}, and the distance to a galaxy is usually measured in the unit of Mpc, a sensible unit for the Hubble constant is: km s^{-1} Mpc^{-1} (i.e. kilometres per second per megaparsec). Note that the "Hubble constant" is in fact rather poorly named, as it is certainly not

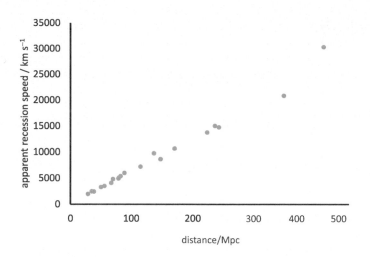

FIGURE 9.3 The Hubble relationship.

constant, but changes in value as the Universe ages. For this reason, it is sometimes referred to as the Hubble parameter, and the phrase "Hubble constant" generally refers to the value that is measured at the present time.

If a graph is plotted of the apparent speed of recession and distance for several hundred clusters of galaxies, the data points will cluster around a straight line. The slope of this line indicates the *rate* at which apparent speed increases with distance. A steeper slope means that apparent speeds increase rapidly as ever more distant galaxies are observed; a shallower slope implies that apparent speeds increase only gradually for distant galaxies. The value of the slope of this line can be measured around 70 km s^{-1} Mpc^{-1}, which means that for every megaparsec of distance out into the Universe, the galaxies and clusters appear to be moving about 70 km s^{-1} faster.

At the time of writing, the most precise measurement of the Hubble constant puts its value at 67.74 km s^{-1} Mpc^{-1} with an uncertainty of about ±0.46 km s^{-1} Mpc^{-1} (i.e. the true value probably lies between about 67.28 km s^{-1} Mpc^{-1} and 68.20 km s^{-1} Mpc^{-1}). By the time you read this, it is possible that an even more precise value for the Hubble constant will have been obtained using one of the new generation of powerful telescopes.

In fact, nowadays, the Hubble relationship is used to *determine* how far away certain objects are. As noted earlier, amongst the most distant objects that astronomers observe are quasars. These are bright, point-like sources of light that lie at the heart of very distant galaxies. They are believed to be super-massive black holes (up to a billion times the mass of the Sun) that are swallowing whole stars that stray too close to them. In the final moments before the stars are engulfed, they are ripped apart and form a swirling disk of material around the black hole. Frictional forces within this disk cause it to get immensely hot and radiate huge amounts of electromagnetic radiation. This is what makes quasars so luminous that they can be seen such a long way away. Astronomers can measure the redshift of a quasar from the absorption or emission lines in its spectrum, infer an apparent speed of motion away from us, and then determine how far away it is using the Hubble relationship. Note, however, that there are a few complications when dealing with "distances" and "speeds" of objects that are so far away, which are addressed in the following sections.

9.4   EXPANDING SPACE

You have seen how cosmologists obtain the observational data on which the Hubble relationship is based. The next task is to look at how it may be interpreted, and then consider the consequences of this interpretation for the properties of the Universe in the distant past.

Wherever cosmologists look, they see galaxies apparently rushing away from the Local Group, and the further away the galaxies are, the faster they appear to move. The interpretation of this is that space *itself* is expanding uniformly, and the same behaviour would be observed *wherever* in the Universe you happened to be.

Here then is the reason why care has been taken up to now to refer to the "apparent speed" of motion of galaxies away from the Local Group. Galaxies are not really rushing away at ever greater speeds as their distance increases; it is simply that the space *in between* the Local Group and the galaxy in question is *expanding*. This expansion of space is what gives rise to the apparent increase in speed of motion with increasing distance.

Now, it is quite difficult to appreciate this for the three-dimensional Universe. So, in order to make things simpler, consider a one-dimensional case, namely a strip of elastic (representing space) with buttons sewn on to it (representing galaxies) (Figure 9.4). Furthermore, imagine that the strip of elastic is being stretched at a constant rate. Showing that uniform expansion of a one-dimensional universe naturally gives rise to the Hubble relationship will hopefully make the idea easier to carry over to the real, three-dimensional case.

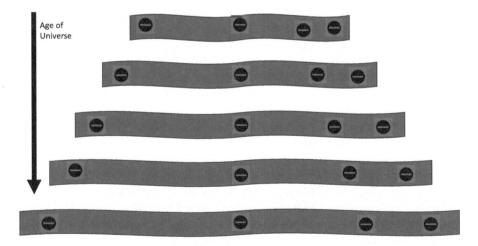

FIGURE 9.4   A one-dimensional expanding Universe.

As the strip of elastic is stretched, so the space that it represents expands uniformly, and all the clusters of galaxies (the buttons) get further apart. As space expands, two clusters of galaxies with a relatively small separation will move apart at a relatively slow rate. However, two clusters of galaxies with a relatively large separation have a greater amount of intervening space between them. So as space expands, there is more space to stretch and they appear to move apart at a greater speed.

Wherever you happen to be in this one-dimensional universe, clusters that are further away from you recede at larger and larger apparent speeds, and this apparent speed depends on the distance, just as described by the Hubble relationship. The Hubble constant is measured with the same value (at the same time) *wherever* you happen to be in this one-dimensional universe, and this is true of the real Universe too.

At times in the past of this one-dimensional universe, clusters of galaxies were closer together, but moving apart at the same constant apparent speed. Because the Hubble "constant" is defined as apparent speed of recession divided by the separation, so the Hubble constant at earlier times was *larger* than it is at the time identified as "today." Conversely, at times in the future of this one-dimensional universe, clusters of galaxies will be further apart, but still moving at the same constant apparent speed. So the Hubble constant at later times will be *smaller* than it is at the time identified as "today."

In the one-dimensional universe model, even though any particular two clusters of galaxies continue travelling apart with the same apparent speed at all times, the rate of expansion of the intervening space becomes progressively smaller. The Hubble constant quantifies the expansion rate of the universe, and clearly the expansion rate of this one-dimensional universe is slowing down. In the real Universe too, the Hubble constant varies with time.

What was demonstrated in one dimension by considering a piece of elastic is also true in the real three-dimensional Universe; it is just a little harder to visualize, and no attempt will be made to do so here.

Having interpreted the Hubble relationship to mean that space is expanding, you can now look at the consequences of this phenomenon. It is believed that the amount of matter in the Universe is constant. So, because the separation between distant galaxies is continually increasing, this implies that the mean density of the Universe – the mass per unit volume – is continually falling. In other words, in the distant past the Universe was very dense, whereas now, the mean density is rather low. This is the first important piece of evidence about the conditions that prevailed in the early Universe.

9.5 STRANGE IDEAS

When talking about the overall structure of the Universe, some rather awkward questions often arise. In the discussion below, these will be addressed in the hope that they will answer any questions you have about the behaviour of the expanding Universe. Before starting though, you should be warned that you will be required to put aside some apparently rational notions of reality and accept a few ideas that may at first seem rather strange. As you will see, it is not only artists and poets who need fertile and wide-ranging imaginations – such characteristics are equally useful for cosmologists. Although some of these ideas require you to put aside common sense, be assured that the mathematical models describing the Universe are based firmly on Einstein's theory of general relativity (which will be discussed later) and are unambiguous.

The first complication is that the Hubble "constant" was larger in the distant past than it is today and, as noted earlier, for this reason it is often referred to as the Hubble parameter. There are two reasons for this. First, as you saw in the one-dimensional universe analogy, even if galaxies keep moving apart at a constant rate, the Hubble constant decreases as time progresses, simply because it is equal to apparent speed divided by separation and the separations keep getting larger. Second, the expansion rate of the Universe may have slowed down or speeded up at different times in its history and this will further change the value of the Hubble constant as time progresses. As a consequence, the distance scale of the Universe in the past is also uncertain.

Also, a word of caution is called for concerning the relationships between red-shift and apparent speed that were mentioned earlier. At high apparent speeds, the physical meaning of "distance" in the Universe needs to be considered carefully. The light that is now observed from rapidly receding galaxies was emitted by them when the Universe was much younger than it is now. In the time it has taken that light to reach the Earth, the Universe has expanded and so distances between galaxies have changed. Therefore, care has to be taken in interpreting the distances and apparent speeds that are measured in an expanding universe.

For these reasons, cosmologists do not usually refer to the apparent speed and distance of rapidly receding, very distant galaxies or quasars, but to their redshift and look-back time. This latter quantity is the time taken for the light emitted by a galaxy or other object to reach Earth and indicates how far back in time the galaxy is seen. Looking at objects far away implies looking back in time, because the light that is detected from them was emitted by the object in the distant past and has taken a substantial amount of time to reach Earth.

Aside from these somewhat technical issues, the idea of an expanding Universe also throws up some rather strange concepts. In response to the claim that space is expanding uniformly, many people ask (not unreasonably): "What is the Universe expanding into?" In fact, the expansion of the Universe is interpreted very differently from an expansion of matter *into* space; rather it is interpreted as an expansion *of space itself*. Space is a property of the Universe, and matter is (more or less) fixed in a space that expands. This was illustrated by the one-dimensional universe analogy considered earlier: the elastic (space) expands uniformly, but the buttons (clusters of galaxies) remain the same size and shape; they are merely carried along by the universal expansion. Similarly, the Earth is not expanding, and nor is the Solar System, the Milky Way, or even the Local Group of galaxies. These objects are all bound together by electric and gravitational forces of attraction between the atoms and molecules of which they are composed. Only beyond the scale of clusters of galaxies does the expansion win.

Although the redshift of distant galaxies has been described as being comparable to a Doppler shift, it is important to realize that there is one vital difference between a "standard" Doppler redshift (such as that caused by speeding ambulances or the random motion of galaxies in the Local Group) and what may be called a cosmological redshift. The Doppler effect is the result of the motion of an object *through* space at a certain speed, whereas cosmological redshifts are caused by the expansion *of* space itself. So, even though a distant cluster of galaxies may have an apparent speed that is (say) 88% of the speed of light, that cluster is not moving rapidly with respect to its local surroundings. In terms of the one-dimensional analogy, the buttons on the strip of elastic are not moving with respect to the local patch of elastic. It is the expansion of space itself that "stretches out" the wavelength of the emitted light as it travels through space. The more space there is between the object emitting the light and the point of observation, the bigger the "stretch," and so the larger the redshift.

Another question that many people ask is: "Where in the Universe is the centre of this expansion?" Well, there is no centre of expansion – all space is expanding at the same rate in all directions, and the same expansion would be measured wherever you happened to be. Perhaps another analogy will help here. Consider the surface of a balloon as representing a two-dimensional universe – one step up in complexity from the strip of elastic considered earlier, but still one dimension short of the real thing – with buttons stuck on the surface representing clusters of galaxies as before. It is only the *surface* of this balloon that represents space – everything inside or outside of the balloon is not part of this universe. As time advances and space expands (i.e. as the balloon is inflated), the clusters of galaxies move further apart with their apparent speeds away from each other increasing with distance. But the centre of expansion (the centre of the balloon) does not lie anywhere within the universe. The real three-dimensional Universe also has no centre of expansion. The problem is that no-one can think in enough dimensions to visualize it properly. This is an important point: do not even attempt to visualize the corresponding situation for the real three-dimensional Universe. It is almost certainly not possible to do so!

Perhaps there is a simple answer to the question: "How big is the Universe?" Well, there is a simple answer, but it is not easy to comprehend. The Universe may be infinite in extent – *and may always have been so.* (The balloon analogy is therefore misleading in this respect, because that describes a universe with finite size.) When it is said that space is expanding, you should *not* interpret this to mean that the overall *size* of the Universe is increasing (if it is infinite, it cannot get any bigger because infinity is the biggest possible!). Rather, you should interpret it to mean that the *separation* of large structures within the Universe is increasing; in other words, galaxies are getting further apart.

A popular misconception is to think of the Universe as originating at a "point in space" and expanding from there. This is quite the wrong visual image, and you should try not to think in these terms. Remember, space is a property of the Universe, not something within which the Universe sits. Furthermore, the current theory for the origin of the Universe implies that the entire infinite space of the Universe, and the raw materials from which the galaxies were built, were all created at the same instant. The separations between objects increase with time, as they are carried along by the expansion of the space that was created at the instant the Universe began. Again, do not even attempt to visualize an infinite, expanding three-dimensional Universe – it is almost certainly impossible for anyone to do so!

A final point is that there is no edge to the Universe either. Because the Universe may be infinite then, by definition, it goes on forever and travelling in a straight line you would never reach an edge. Even if the Universe were finite in size though, you would never reach an edge. Travelling in a straight line in a finite Universe, you would eventually end up back where you started, just as an ant would crawling over the surface of the two-dimensional universe model represented by the surface of a balloon.

The preceding few paragraphs provide a rather mind-bending excursion for most people! The problem is that humans are only used to comprehending things on a much smaller scale of time and space than is necessary to grasp properly the immensity of the Universe. The ideas can be expressed mathematically but would be an unnecessary and lengthy detour from the main story. Nevertheless, the basic ideas are not so difficult if you are prepared to discard some ideas that are "common sense" in everyday experience. To summarize:

- The Universe is probably infinite, with no centre and no edge, and it was probably always infinite, as far back in time as theories take us.
- It makes no sense to ask what is outside the Universe because space is a property of the Universe itself and does not exist elsewhere.
- Space itself is expanding uniformly such that the separation between distant galaxies increases with time, and the overall density of the Universe decreases.
- A consequence of this uniform expansion is that the redshift of distant galaxies increases with increasing distance from the place of measurement.

9.6   THE AGE OF THE UNIVERSE

For the time being, follow this first big clue of cosmology: space is expanding. You can therefore conclude that, in the past, all the galaxies were closer together than they are now. If the assumption is made that all the galaxies that can be observed have been moving at their present apparent speeds since the Universe began, then the Hubble constant can be used to calculate a rough age for the Universe.

If the Hubble constant is about 70 km s^{-1} Mpc^{-1}, then two galaxies that are currently 500 Mpc apart have an apparent speed between them of 70 km s^{-1} Mpc^{-1} × 500 Mpc = 35,000 km s^{-1}. (Remember, that is 70 km s^{-1} faster for every Mpc further apart.) So, if you imagine "running the film backwards," these two galaxies would have been "zero" distance apart at a certain time in the past. That time is given by 500 Mpc / 35,000 km s^{-1} because the time for a "journey" is simply given by the distance travelled divided by the speed. But distance divided by speed is just 1/Hubble constant, so the quantity "1/Hubble constant" provides a rough value for the age of the Universe, assuming that the expansion rate has been constant since time began.

So what is the age of the Universe? As noted above, an approximate value for the Hubble constant is 70 km s^{-1} Mpc^{-1}. To work out an age for the Universe, the units need to be rationalized. At the moment, the Hubble constant has a unit that includes two different measures of distance: megaparsecs and kilometres. As noted earlier, 1 megaparsec is equal to 3.1×10^{19} kilometres, so the value of the Hubble constant can be written as: Hubble constant = (70 km s^{-1} Mpc^{-1})/(3.1×10^{19} km Mpc^{-1}) = 2.26×10^{-18} s^{-1}. An estimate for the age of the Universe is then $1/(2.26 \times 10^{-18}$ s$^{-1}) = 4.42 \times 10^{17}$ s, or equivalently around 14 billion years. In practice, the true age of the Universe will be slightly different from this value because the expansion of the Universe has *not* proceeded at a constant rate since time began.

10 The Cooling Universe

Having looked closely at the expansion of the Universe, this chapter examines the second major piece of evidence for an evolving Universe, namely the observation that the Universe is gradually cooling. If the Universe is cooling, then it must have a temperature, and that may seem a rather strange concept. What is the temperature of the Universe? After all, nowadays the Universe largely consists of almost empty space between the galaxies. However, space is not as empty as you might suppose. Even away from stars and galaxies, space still contains electromagnetic radiation. On average, every cubic metre of space contains about 400 million photons! These photons constitute the "heat radiation" of the Universe, and their spectrum corresponds to a particular temperature.

10.1 BLACK-BODY RADIATION

To begin, consider the question, what *is* temperature? Roughly speaking, as an object becomes hotter, the atoms and molecules of which it is composed move around or vibrate more rapidly. As an object cools down, its atoms and molecules move around or vibrate ever more slowly. There will come a point when the atoms and molecules cease moving altogether, and such an object would then be at the absolute coldest temperature possible. Because there is such a thing as an absolute minimum temperature, in many areas of science, the most useful way to describe temperatures is to use the absolute temperature scale. This is also known as the Kelvin temperature scale, in honour of the British physicist William Thomson, Lord Kelvin who developed the idea in 1848. On this scale, zero kelvin (0 K) is the coldest temperature possible.

Zero kelvin corresponds to about −273°C on the more familiar Celsius scale, and a temperature interval of 1 K is identical to an interval of 1°C. A comparison between the Kelvin and Celsius temperature scales reveals that the melting temperature of water ice (0°C) is equivalent to about 273 K and the boiling temperature of water (100°C) is equivalent to about 373 K.

Now, as you saw in earlier, when a hot-plate on an electric cooker is heated up, it glows red-hot and radiates energy in the form of photons. These photons have a range of energies, but the precise distribution of photons – the relative numbers that are emitted with any particular energy – depends on the temperature of the hot-plate. As you know, the distribution of photons plotted against photon energy is simply the spectrum of the radiation. It turns out, quite reasonably, that as the temperature of an object is increased, it emits photons of progressively higher energies. There will still be a distribution of photons with different energies (i.e. a spectrum), but the *mean* photon energy will shift to higher values. An object whose emission has a mean photon energy in the blue part of the spectrum (about 3 eV) will be hotter than one whose emission has a mean photon energy in the red part of the spectrum (about 2 eV), for instance. But there is no need to restrict this relationship between photon energy and temperature to merely the visible part of the electromagnetic spectrum. At higher energies (shorter wavelengths) than blue light there are the ultraviolet and X-ray regions. Objects whose emissions have mean photon energies in these parts of the spectrum must be extremely hot. Conversely, at lower energies (longer wavelengths) than red light are the infrared and microwave regions. Emissions that have mean photon energies in these ranges would indicate much cooler temperatures.

As a rough rule of thumb, photons with a mean energy of about 1 eV will be produced by a body at a temperature of around 3000 K, and the mean energy of the photons emitted by other bodies is proportional to the temperature. The continuous spectral distributions of many objects have precisely the same shape, they are merely shifted to different energies or wavelengths, and each is what is known as a black-body spectrum or thermal spectrum.

From everyday experience, you may be aware that a black surface absorbs more radiation than a silver surface. In fact, a perfectly black surface will absorb all the radiation that falls upon it and, in a steady-state (or equilibrium) situation where it remains at a constant temperature, it will also emit all this radiation back again. (This assumes that the surface is in a vacuum and cannot exchange energy with its surroundings in any other way.) The phrase "black body" is therefore used as a short-hand to describe any object that behaves as a perfect absorber and emitter of radiation. The crucial features of black-body spectra are that they all have the same continuous shape, they contain no emission or absorption lines, and the mean photon energy (or corresponding wavelength) depends *only* on the temperature of the object (Figure 10.1).

The key to understanding how black-body spectra are produced is that the object and the radiation are in thermal equilibrium. As much radiation is being absorbed as is being emitted at every instant, and the object therefore remains at a constant temperature. It can therefore be said that the radiation also possesses this same temperature. The conditions necessary to create such a situation are generally those of high temperatures and large amounts of energy. Under such conditions, photons are rapidly absorbed and re-emitted by matter.

As you know, in order for a photon to be absorbed by an atom, there must be a pair of energy levels whose separation is equal to the energy of the photon concerned. To guarantee that a photon of virtually *any* energy could be absorbed, there must be a great many energy levels, very close together, and extending over a large range of energy. It turns out that this situation will exist when atoms are ionized (i.e. when one or more electrons are removed from an atom), and also when atoms are arranged in a metal. So, in both of these cases, photons of *all* energies may be absorbed and emitted, and a continuous spectrum is produced. The continuous spectrum of the Sun and other stars (ignoring the absorption lines superimposed on top) may be approximated by black-body spectra.

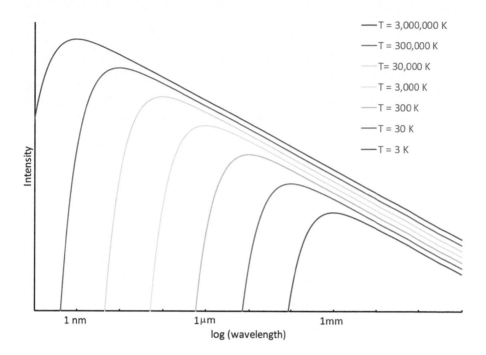

FIGURE 10.1   Black-body spectra corresponding to various temperatures.

10.2   THE COSMIC MICROWAVE BACKGROUND

Objects in the Universe emit electromagnetic radiation across the whole spectrum from radio waves through infrared, visible radiation, ultraviolet, and X-rays to gamma-rays. The spectra from individual objects are, in some cases, characteristic of thermal processes, and so have the continuous black-body shape. If astronomers observe the Universe in the microwave part of the spectrum – that is at wavelengths of a few millimetres – a remarkable phenomenon is observed. What they detect is a background microwave "glow" coming from the whole Universe, and wherever they look (away from individual stars or galaxies) it has virtually the same spectral distribution and intensity. Moreover, the shape of the spectrum is that of a thermal or black-body source. This cosmic microwave background (CMB) was discovered in 1964 by US astronomers Arno Penzias and Robert Wilson.

At the time they were testing new microwave technology for tracking satellites at the Bell Telephone Laboratories. They unexpectedly detected a microwave glow from the entire sky. At first, they attributed this to interference from pigeon droppings in the horn antenna they were using, but after cleaning the detector and disposing of the pigeons nesting within it, they concluded that it was a real signal.

Over the years since Penzias and Wilson first discovered the CMB, its spectrum and variation across the sky have been investigated and mapped with increasing precision. The first accurate measurements were those made using the Cosmic Background Explorer (COBE) satellite, which was launched in 1989. The average spectrum of the radiation, over the whole sky, measured by COBE is an excellent fit with the theoretical curve that would be expected from a black-body source at a temperature of 2.73 K. Sometimes, this value is quoted approximately as simply 3 K (i.e. $-270°C$). This has been refined even further with the Wilkinson Microwave Anisotropy Probe (WMAP) satellite launched in 2001, which measured fluctuations in temperature across the sky at a level of only 5×10^{-5} K (Figure 10.2).

In precise physical terms, the temperature of the CMB radiation is now (coincidentally) one hundred times colder than the normal melting temperature of ice (i.e. 273 K). Yet, the spectrum has exactly the same shape as is observed in, say, a furnace whose walls are a thousand times hotter at a temperature of 3000 K, where interactions between radiation and matter rapidly create the stable distribution of photon energies necessary to produce a black-body spectrum. Three thousand kelvin can be thought of as roughly the *minimum* temperature at which atoms and radiation can interact significantly (corresponding to a mean photon energy of about 1 eV). Below this temperature, the energy levels in hydrogen atoms are simply too far apart to absorb many of the photons corresponding to a black-body spectrum.

At a temperature of only about 3 K, a steady state, or thermal equilibrium between matter and radiation in the Universe is virtually impossible to establish, because the energy of most of the photons is so very small when compared with the separation of energy levels in hydrogen atoms. So how can radiation that is now far too cold to interact with matter, to any significant extent, have acquired a thermal spectrum, when thermal spectra are generally characteristic of processes at least a thousand times hotter? If it were not for the previous observation of an expanding Universe, there would be a real puzzle here.

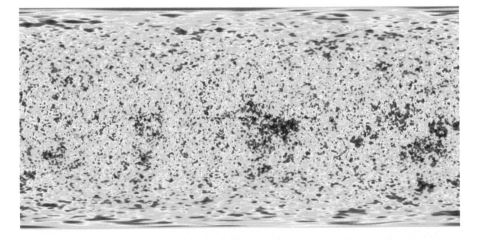

FIGURE 10.2 Fluctuations in the CMB radiation. Red regions are warmer and blue regions are colder than average by 0.0002 degrees. (Credit: NASA.)

The solution is that the CMB radiation now seen was emitted by matter at a time in the distant past when the Universe was much hotter than it is today. As the CMB radiation travels through space from distant parts of the Universe, so the wavelength of the radiation is redshifted by the expansion of the intervening space. Looking back to the time when the CMB radiation and matter were last in equilibrium with each other entails a redshift of about 1000. In other words, the wavelength corresponding to the mean photon energy of the CMB radiation today (about 2 millimetres) was 1000 times shorter (i.e. it was about 2 micrometres) when the radiation was emitted. The wavelength of the radiation has been "stretched out" by the expansion of the Universe. A lengthening of the wavelength from 2 micrometres to 2 millimetres corresponds to a reduction in photon energy from 1 eV to 0.001 eV. So the expansion of the Universe has shifted the mean photon energy of the radiation from the infrared part of the spectrum down into the microwave region where it is observed today.

The CMB radiation is therefore a relic of the time when radiation and matter in the Universe existed in equilibrium, at a temperature of around 3000 K. The CMB photons pervade the entire Universe, wherever astronomers look. This indicates that the entire Universe was once at a much higher temperature.

10.3 THE HOT BIG BANG

The expansion of the Universe (as expressed by the Hubble relationship) shows that objects in the Universe were closer together in the past than they are today. In other words, the Universe was *denser* in the past than it is now. The cosmic microwave background radiation shows that the Universe was *hotter* in the past than it is now. These are the two main pieces of evidence pointing to the fact that the Universe originated in what has become known as the Big Bang.

In fact, the idea of a Big Bang actually pre-dates not only the discovery of the cosmic microwave background, but also Hubble's discovery of the expansion of space. In 1927, the Belgian priest Georges Lemaître had already proposed that the Universe began with an "explosion of a primeval atom." The idea was subsequently developed by the Russian-American physicist George Gamow who predicted the existence of the cosmic microwave background radiation, long before Penzias and Wilson discovered it. However, the name "Big Bang" was not coined until 1949 when the British astronomer Fred Hoyle used the term in a BBC radio programme in order to denigrate the theory, which he disliked and which was a rival to his own "steady-state" model of the Universe.

The now standard model for the origin and early evolution of the Universe is sometimes known as the hot Big Bang model. Space and time were created in this event, and space has expanded as time has progressed ever since. The story of the evolution of the Universe from the time of the Big Bang to the present day will be presented in Chapter 16. However, in order to discuss such an immense topic, it is necessary to appreciate the role that four fundamental interactions each play in the evolution of the Universe. In Part III, you will learn about these four interactions in turn, and you will discover how attempts are being made to unify these into a single, coherent, theory of everything.

Part III

Universal Processes

In Chapter 16, you'll get a more complete picture of what the Universe was like when the cosmic microwave background radiation last interacted with matter, which was long before the first galaxies and stars were formed. It's even possible to describe times that are much earlier than this, when the atoms, nuclei, leptons, and quarks came into existence. First though, it's necessary to summarize what's known about the interactions of matter and radiation, from experiments that have been performed here on Earth. These in turn suggest some things that might occur at energies that are higher than those which can be achieved in Earth-bound laboratories, but which were very common in the early Universe.

This involves continuing to investigate the structure of atoms and shows how cosmology and particle physics are inextricably linked together. Cosmologists have learned things from studying the cosmic microwave background radiation that makes particle physicists come up with new ideas about the interactions of particles at high energies. Likewise, particle physicists make predictions based on theories developed to explain laboratory results and then look at the wider Universe for evidence to test them against.

In Chapter 16, you'll see that information from particle physics is needed to explain the history of the Universe. To do this, we need a quantum theory that describes the interactions between particles, and an understanding of how these interactions change when the particles involved interact with high kinetic energies. It turns out that *all* possible interactions can be explained in terms of four fundamental interactions, namely:

1 Electromagnetic interactions are what give rise to the electric forces between charged particles (such as electrons and protons in atoms), the magnetic forces caused by moving electric charges (such as electric currents), and the emission and absorption of electromagnetic radiation (such as light). A residual electromagnetic interaction allows atoms to bind together to make molecules and so is responsible for all chemical, and ultimately biological, processes.

2 Strong interactions are responsible for the very strong forces between quarks *inside* protons and neutrons. A small left-over effect of the strong interactions between quarks allows protons and neutrons to bind together to form atomic nuclei.

3 Weak interactions give rise to processes such as radioactive beta-decay in which both quarks *and* leptons are involved. In particular, they are crucial for processes in which particles *transform* from one type to another.

4 Gravitational interactions are the familiar forces that make apples fall to the ground, keep planets orbiting around stars, and control the rate at which the Universe is expanding. However, gravity has essentially no effect at all within atoms. Nonetheless, when matter is accumulated into enormous, electrically neutral bodies, like stars and planets, it is the force of gravity that dominates everything else.

These four interactions are the subjects of Chapters 11–14 which will look in turn at the way in which each of these interactions operates and at how their strengths compare. In Chapter 15, you'll see that, in certain circumstances, these interactions are probably not as distinct as you might have thought. It turns out that they may all be different aspects of a more unified description of nature. While they may appear as different as (say) ice, water, and steam, just like these three substances which are all composed of H_2O molecules, the different interactions too have underlying similarities.

Remember, the aim is to understand the Universe. To do this you need to know about the different stages involved in building up nuclei, atoms, planets, stars, and galaxies out of the material that emerged from the Big Bang. The key to all these processes is an understanding of the four fundamental interactions that govern the behaviour of the contents of the Universe. It is in this third part of the book, therefore, that the second big question posed in Chapter 1 is addressed – what rules does the Universe follow?

11 Electromagnetic Interactions

The behaviour of particles within atoms and molecules involves the interactions between electrically charged particles. The most obvious of these is the interaction between electrons and protons in an atom, but they are also responsible for the behaviour of molecules too. In fact, electromagnetic interactions have two distinct aspects: we recognize electric forces between electrically charged objects and we also observe magnetic forces between moving electric charges (which we call electric currents). Furthermore, electromagnetic radiation (such as light) is emitted or absorbed (as photons) in processes involving these forces. These three apparently distinct features of electromagnetic interactions will now feature in the rest of this chapter. The intention here is to understand their unification, described by the theory of quantum electrodynamics (QED).

11.1 ELECTRIC AND MAGNETIC FORCES

The law that describes the force of electrical attraction and repulsion between charged particles was discovered by the French physicist Charles Augustin de Coulomb in 1785 and can be expressed in the following way:

> Coulomb's law: two stationary particles of unlike (or like) charge, attract (or repel) each other with an electric force that is proportional to the product of their electric charges divided by the square of their separation.

Coulomb's law is therefore another example of an inverse square law, just like the relationship between brightness and luminosity introduced in Chapter 8. The law also says that if one particle has a positive charge and the other has a negative charge (i.e. unlike charges), then there is a positive force of attraction. Conversely, if the particles have charges with the same sign (either both positive or both negative), the force between them is repulsive. In other words, unlike charges attract each other, and like charges repel each other.

Coulomb's law may be used to investigate the strength of attraction between particles such as protons and electrons. The result of such a calculation is remarkable. The electric force between a proton and an electron at the typical separation that applies within a hydrogen atom is equivalent to about one-hundredth of the weight of a grain of sand. Yet a grain of sand contains around one hundred billion billion atoms. Now, the weight of an object is a measure of the gravitational force of attraction between the Earth and the object in question. So, the electric force between a single proton and electron is about the same strength as the gravitational force between the Earth and a billion billion atoms! From this, you can see that the gravitational force exerted by the Earth on a *single* atom is utterly negligible compared with the electric forces within it.

Magnetism may be most familiar to you from the interaction of magnetic materials, such as a bar magnet and a compass needle. However, the magnetic properties of metals derive from the motion of the electrons that they contain. Nowadays, perhaps the commonest source of "moving electrons" is any electrical apparatus, because an electric current is simply that: a flow of electrons.

Around 1820, the Danish scientist Hans Christian Oersted noticed that an electric current flowing in a wire affects a compass needle placed close to it, so demonstrating that the movement of charge produces a magnetic force, registered by the magnetic material in the compass needle. The converse effect is that a bar magnet exerts a magnetic force on moving charges. Around the same time, the French scientist André-Marie Ampère showed that if two wires close to each other carry currents that flow in the same direction, the wires attract each other with a magnetic force, whereas if the currents flow in opposite directions, there is a magnetic repulsion. Oersted and Ampère had therefore laid the foundations for understanding magnetic forces, just as Coulomb had done for electric forces.

11.2   ELECTROMAGNETIC FIELDS AND RADIATION

In 1864, at a meeting of the Royal Society, James Clerk Maxwell presented a set of equations that unified the laws of magnetism with those of electricity. His subsequent book outlining the theory, *A Treatise on Electricity and Magnetism*, published in 1873, explained all of the then-known effects of electromagnetism, with the distinctive prediction that light is a form of electromagnetic radiation.

The key concept to Maxwell was that of an electromagnetic *field*. Coulomb's law had described the electric force exerted by one charge on another charge, when the two are positioned a certain distance apart. Now suppose that the second charge is removed and replaced at the same location by a charge that is ten times larger. How does the force on this charge compare with that which was exerted on the original charge? Coulomb's law shows that the force has the same direction but is ten times larger. Maxwell's perspective was that a charge produces an electric field, spread out over the whole of space, *whether or not* there happens to be another charged particle responding to it at any particular point in space, at any particular time (Figure 11.1). The force becomes apparent only when the second charge is placed in the field of the first.

The electric force that acts on a charge placed in an electric field is simply the strength of the electric field due to the first charge at that point in space multiplied by the value of the second charge. Electric field strength can be expressed in the standard unit of newtons per coulomb, N C^{-1}. Notice that the second charge introduced into the field of the first will also produce its own electric field and give rise to a force acting on the first charge due to the electric field of the second. The two forces will be equal in magnitude, but act in opposite directions.

You may find the field idea to be an unnecessary complication, if all that results is Coulomb's original law of force. However, magnetic fields are much easier to work

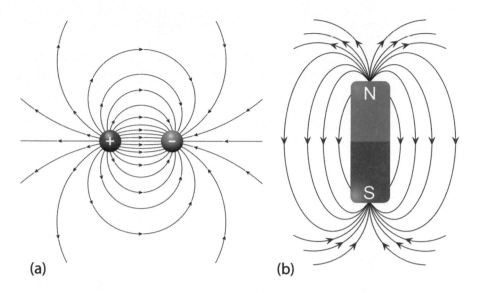

(a)   (b)

FIGURE 11.1   (a) Electric fields and (b) magnetic fields.

with than magnetic forces and the magnetic fields due to currents flowing in wires
are much easier to visualize. Note that, unlike electric field lines, magnetic field lines
are always "closed" loops.

So, Maxwell's concept of electric and magnetic fields was initially useful for two
reasons. It meant that:

- Coulomb's law could be reinterpreted as implying that a *stationary* electric
  charge gives rise to an *electric* field.
- Oersted's and Ampère's discoveries could be reinterpreted as implying that
  a *moving* electric charge (i.e. an electric current) gives rise to a *magnetic*
  field.

The next discovery in this area had been made around 1830, independently by a pair
of scientists, Englishman Michael Faraday and the American Joseph Henry. Their
observation could be interpreted as showing that a *changing* magnetic field produces
an electric field. This phenomenon is called electromagnetic induction and is the
basis of a dynamo.

Maxwell's great contribution was to predict one more phenomenon: that a *changing* electric field produces a magnetic field. The previous three parts of electromagnetic theory had been developed in response to experiments: the electric field of
a stationary charge came from observations by Coulomb; the magnetic field of a
steady current from studies by Oersted and Ampère; and the electric field produced
by a changing magnetic field from measurements by Faraday and Henry. There was
no data that required Maxwell's fourth idea, which was that a magnetic field can be
produced by a changing electric field. Instead the prediction was driven by a study of

the equations describing the previous three effects, and by Maxwell's sense that the addition made a more "elegant" set of equations. The four differential equations that we now use to encapsulate Maxwell's ideas are indeed a remarkably symmetrical, elegant, and concise set of statements. They are referred to as Maxwell's equations of electromagnetism.

However, scientists must make predictions, or illuminate old problems, if their contributions are to have a lasting impact. Maxwell's prediction was something entirely new: according to his equations, because a changing magnetic field generates an electric field, and a changing electric field generates a magnetic field, an electromagnetic wave could be set up, with electric and magnetic fields at right angles to each other and to the direction in which energy is transported (Figure 11.2). Furthermore, these waves would all travel through a vacuum (i.e. empty space) with the universal (constant) speed: 3.00×10^5 km s^{-1}, which is also the speed of light. The speed of light is simply related to the electrical and magnetic properties of a vacuum. These are characterized respectively by the so-called permittivity of empty space and the permeability of empty space.

This showed not only that visible light is a form of electromagnetic radiation, but that electromagnetic radiation exists with a vast range of possible wavelengths. All the various types of radiation from radio waves and microwaves, through infra-red, visible, and ultraviolet radiation, to X-rays and gamma-rays are merely different manifestations of the same oscillating electric and magnetic fields described by

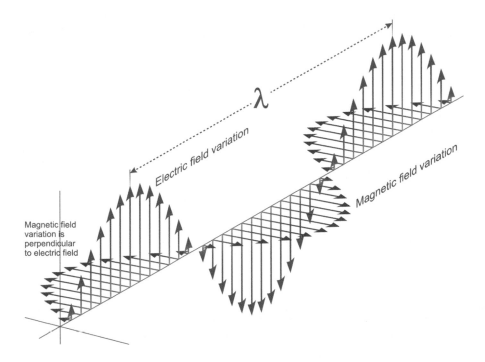

FIGURE 11.2   Electromagnetic waves.

Maxwell. It is hard to imagine more far-reaching consequences of trying to tidy up a set of equations. Most of the technology that is vital to modern society depends on the understanding of electricity, magnetism, and electromagnetic radiation, and all of these devices have been enabled by the insight that Maxwell provided.

11.3   QUANTUM ELECTRODYNAMICS

Maxwell completed his theory of electromagnetism in 1873, and you may be wondering whether that was the last word on this phenomenon. In fact, it was not by a long way. The quantum physics of atoms is inextricably linked with the emission and absorption of electromagnetic radiation (photons). So clearly there was a need to unite Maxwell's theory of electromagnetism with the quantum model of the atom if a correct description of atoms and radiation was to be obtained. As you read earlier, another important development in physics in the early part of the 20th century was Einstein's theory of special relativity, published in 1905. A key result of this theory is that the kinetic energies of particles travelling at a substantial fraction of the speed of light do not obey the conventional formula that applies in everyday situations. Electrons in atoms have a range of possible speeds, and the most probable speed for the electron in a hydrogen atom is around 1% of the speed of light. Even at this speed, the conventional formula for kinetic energy is in error by 0.005%. At higher speeds (which are possible in atoms), even larger errors occur if special relativity is not taken into account.

A fully consistent explanation of the properties of atoms, electrons, and radiation must therefore combine *electromagnetism* with *quantum physics* and *special relativity*, to produce what is called a relativistic quantum theory of these properties. The first stage in this process was completed by the English theoretical physicist Paul Dirac in 1928. To achieve a high precision in describing tiny corrections to the quantum model of the atom as a result of special relativity, Dirac predicted the existence of a new particle, with the same mass as the electron, but with positive charge. As you saw in Section 6.3, this particle is called the positron, or antielectron, and given the symbol e$^+$. Now, as you've also seen earlier, in the process of pair creation, high-energy photons can create electron–positron pairs, thereby producing new particles from purely electromagnetic energy. The mass energy of an electron or positron is just over 500 keV, so to create an electron–positron pair requires more than 1 MeV of energy.

Pair creation, therefore, requires photon energies that are about a million times greater than the energies of a few eV that are involved in the energy levels of the hydrogen atom or photons of visible light! So, at first sight, there is no need to consider positrons in atomic physics. After all, the idea of energy conservation is basic to physics. Why consider the creation of positrons, when none ever emerge from atoms, apart from the comparatively rare unstable nuclei that undergo beta-decay?

It turns out that there *is* a need to consider positrons: the conservation of energy is something that has to work out only on long timescales; on much shorter timescales energy accounting may be "relaxed," provided the accounts are settled in the long run. The basic rule for this had been stated by Werner Heisenberg in the

1920s: a failure of energy conservation may be tolerated for a short time provided that the energy deficit multiplied by the time interval is less than the value of the Planck constant divided by 4π. After the time limit specified, the energy debt must be made good. In fact, this is just another form of the Heisenberg uncertainty principle that you met in Chapter 4 in relation to how precisely positions and velocities of electrons in atoms can be known. This energy–time uncertainty principle embodies an important feature of the quantum world: whatever is allowed to happen will do so, sooner or later.

The uncertainty principle is a bit like an arrangement with a rather eccentric loan company who tell you that may borrow £100 for up to 100 days, or £1000 for up to 10 days, or £10,000 for no more than a single day. In this scenario, anything is allowed, as long as the debt multiplied by its duration is no more than 10,000 pound-days.

In physics, the general feature of the energy–time uncertainty principle is that you may push it to the limit: any failure of the conservation of energy may be allowed for a certain time, as long as the limit above is not exceeded. The terms are not generous, but they are totally flexible, within the specified credit limit.

The energy–time uncertainty principle led to the idea that empty space, even inside an atom, is not really empty. In informal terms, it is as if electrons and positrons were constantly appearing out of nothing, and then disappearing before their credit has run out. The space inside a hydrogen atom (and inside any other atom) is filled with transient electron–positron pairs. The (negatively charged) electrons that are created are drawn towards the (positively charged) proton at the centre of the hydrogen atom. This effectively "hides" some of the charge of the proton, in a process referred to as screening, which you read about in Chapter 5 in the context of multi-electron atoms.

If you were to measure the electric force of attraction on an electron produced by the nucleus of a hydrogen atom (i.e. a proton) when situated relatively far away from it, you would get an answer that agrees with Coulomb's law. This equation describes the effective strength of electromagnetic interactions when screening due to transient electron–positron pairs *is included*. However, if you were to measure the same effect when much closer to the nucleus, some of the transient electron–positron pairs would be further away. Consequently, there would be *less* screening, so the effective charge of the proton would appear slightly larger, and the electric force would consequently increase. In other words, at small distances, the electromagnetic interaction will appear *stronger* than it does at larger distances.

These quantum effects modify the electric force in a hydrogen atom only over distances that are about 1% of the typical separation of an electron and proton. The result is to modify the energy levels by a fractional amount that is only about one part in a million. Yet the discovery and explanation of such tiny effects led to the development of a whole new theory, called quantum electrodynamics (QED). Quantum electrodynamics is the most complete theory of electric and magnetic interactions known. It was developed in the 1940s by Richard Feynman, Julian Schwinger, and Sin-Itiro Tomonaga. QED incorporates descriptions of the emission and absorption of photons and is needed to understand many features of the subatomic world. In this theory, all electric and magnetic forces are envisaged as arising from the *exchange* of photons between charged particles. Electricity, magnetism, electromagnetic radiation, and the behaviour of electrons in atoms are merely different aspects of the same phenomenon. Many confirmations of this theory have now been obtained.

The final surprise of QED has already been hinted at above. It concerns how the strength of the electromagnetic interaction varies depending on the energy at which the effect is investigated. When atoms are probed in high-energy experiments, distances close to the nucleus are investigated. In these regions, the strength of electromagnetic interactions is greater than the strength predicted by Coulomb's law, because screening of the nucleus by transient electron–positron pairs is less effective. In other words, electromagnetic interactions appear *stronger* when they are investigated at high energies. In Chapter 15, you will see that the varying strength of fundamental interactions with energy is a crucial feature that provides an understanding of the conditions that prevailed in the early Universe.

12 Strong Interactions

The force that binds quarks together inside nucleons (i.e. neutrons and protons) is known as the strong interaction and has a very short range. It operates essentially only within the size of a nucleon. The strong interaction is what allows two up quarks and a down quark to bind together to form a proton, or two down quarks and an up quark to bind together to form a neutron. However, there is a residual strong interaction *between* nucleons, which you can imagine as "leaking out" of the individual protons and neutrons. This left-over force is enough to bind them together in nuclei and is similar in nature to the residual electromagnetic interactions between atoms that are responsible for the formation of molecules.

This chapter examines the first of two fundamental interactions that are likely to be the least familiar to you. The strong interaction is vital for understanding the Universe because it is what allows protons and neutrons to exist and also what allows nuclei consisting of protons and neutrons to exist. By understanding how the interaction operates, you will be able to appreciate the vital role it played in the very early Universe. Quite simply, the strong interaction enabled the fundamental particles created out of the Big Bang to begin to bind together to form the more familiar nuclei of which the present-day Universe is composed.

12.1   QUARKS AND GLUONS

It is amazing just how strong the strong interaction between quarks is. At a separation of around 10^{-15} m – the typical size of a proton or neutron – the force of attraction between a pair of quarks is equivalent to the weight of a couple of elephants. As might be suggested by its name, the strong force of attraction between two up quarks is much larger than the electric force of repulsion between them. It is this strong force that prevents quarks from being liberated in high-energy collisions. Free quarks are *never* seen to emerge from such processes: quarks only exist confined within baryons or mesons. It is as if they are stuck together with very strong glue.

Consider, for example, an experiment that was conducted at the Large Electron–Positron (LEP) collider at the European Organization for Nuclear Research (CERN), near Geneva. In this experiment, positrons of kinetic energy 50 GeV collided with electrons travelling in the opposite direction, with the same kinetic energy. As a result, an electron and a positron may annihilate each other, producing a quark–antiquark pair, with total energy 100 GeV or 10^{11} eV. These energies are certainly impressive: 10^{11} times greater than the few eV that is typical for the energy levels of the hydrogen atom and even 10^5 times greater than the combined mass-energy of the electron and positron (i.e. 1 MeV). But what can such high energies achieve, against the strong force? In fact, 100 GeV of energy can separate a quark–antiquark pair by only about 100 times the size of a proton.

But if quarks do not emerge from these collisions, what happens to the 100 GeV that is put into the collision? Energy cannot be destroyed but can only be transformed from one kind to another. In the mess of debris that results from high-energy collisions between electrons and positrons, there is a tell-tale clue as to the original interaction that occurred at very short distances: it often happens that hadrons emerge as a pair of jets, with each jet made up of a number of hadrons (Figure 12.1). (Recall from Chapter 7 that hadrons are composite particles made of quarks and antiquarks.)

The basic interaction that produces the pair of jets is as follows. First, the electron (e⁻) and positron (e⁺) annihilate each other and can be thought of as producing what is known as a virtual photon. The reason for this name is that the virtual photon only has a temporary existence and immediately undergoes a pair-creation event, giving rise to a quark–antiquark pair. Both the matter–antimatter annihilation event and the pair-creation event are electromagnetic interactions. The virtual photon is understood merely as a way of carrying the electromagnetic force from the annihilation event to the pair-creation event; it does not escape from the process and is never detectable. Similarly, the quarks and antiquarks created from the virtual photon cannot separate indefinitely. Instead, their kinetic energy and mass energy are converted into the kinetic energy and mass energy of many more matter and antimatter particles, including lots more quarks and antiquarks. Remember, as long as energy is conserved, any energy transformations are possible. The many quarks and antiquarks then combine to form a variety of hadrons, and it is only the hadrons that then emerge from the collision as a pair of jets.

It turns out that this is not the whole story of what may occur in an electron–positron collision producing hadron jets: sometimes *three* jets are produced. This process involves a quite new ingredient, called a gluon. Just as photons are the quanta of energy associated with electromagnetic interactions, so gluons are the quanta of energy whose emission and absorption are regarded as the origin of strong

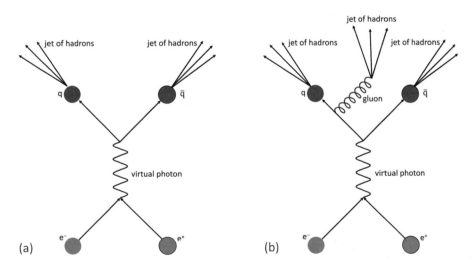

FIGURE 12.1   Jets in strong interactions: (a) two jets of hadrons, (b) three jets of hadrons.

interactions. They are responsible for "gluing" the quarks strongly together inside hadrons. However, unlike photons, but like quarks and antiquarks, gluons cannot escape to large distances. Nonetheless, a quark (or indeed an antiquark) may emit a gluon and in doing so the quark loses some energy. But the quark still has plenty of energy left and can go on to produce a jet of hadrons, just as already described above, and a second jet of hadrons is produced by the antiquark. In addition, the mass energy and kinetic energy of the gluon is quickly turned into the mass energy and kinetic energy of further pairs of quarks and antiquarks. These in turn combine with each other to form various hadrons, and the hadrons produced from the gluon then give rise to a *third* jet emerging from the process.

One way of characterizing the strength of strong interactions is to measure how often a three-jet outcome occurs, when compared with a two-jet outcome, because a three-jet process is the one that involves a gluon. At an interaction energy of 100 GeV, there is roughly one three-jet outcome for every ten two-jet outcomes. Furthermore, it turns out that as the energy of interaction *increases*, the fraction of three-jet outcomes *decreases*. That is, there are progressively fewer outcomes in which a gluon is involved, as the energy of interaction between the colliding electron and positron is turned up. Therefore, the strength of strong interactions must decrease at higher energies.

So, whereas in QED the strength of electromagnetic interactions *increases* with increasing energy, the strength of strong interactions *decreases* with increasing energy.

By comparing the types of collision that involve strong interactions with those that only involve electromagnetic interactions, it is also possible to directly compare the strengths of these two interactions. At an interaction energy of 100 GeV, it has been calculated that the strength of strong interactions is about ten times greater than the strength of electromagnetic interactions.

This may be a good point to step back for a moment and remind you *why* you are being told about the details of these processes that rely on the strong interaction. Remember, the goal is to understand the Universe, and in particular what happened in the very early Universe, soon after the Big Bang. At those very early times, the Universe was very hot and dense, so high-energy reactions, such as those just described, played a very important part in determining how the Universe evolved. This summary of strong interactions will be completed by describing the theory that explains how these interactions actually work.

12.2   QUANTUM CHROMODYNAMICS

The quantum theory of the strong interactions between quarks and gluons is called quantum chromodynamics (QCD); it was developed in the 1970s by Harald Fritzsch, Heinrich Leutwyler, and Murray Gell-Mann. It doesn't deal with force directly, but instead with the interactions between particles and quanta. In QED the main interaction is between electrically charged particles and photons. In QCD there are two basic interactions: quarks (and antiquarks) interact with gluons, and gluons also interact with themselves. It is interactions *between* gluons that are responsible for the fact that the strength of strong interactions decreases with increasing energy.

You may be wondering why this theory is called quantum *chromo*dynamics, as you may be aware that "chromo" comes from the Greek word for "colour." In fact, the interactions between quarks and gluons are described in terms of a new property of matter that is known as colour charge, by analogy with conventional electric charge. Just as electromagnetic interactions result from "forces" between electrically charged particles, so strong interactions result from "forces" between colour-charged particles. However, whereas conventional electric charge comes in only one type that can either be positive or negative, colour charge comes in *three* types, *each* of which can be "positive" or "negative." These three types of colour charge are known as red, green, and blue, and their opposites are antired (or the colour cyan), antigreen (or the colour magenta), and antiblue (or the colour yellow). It is important to note that colour charge has *nothing* to do with colours of light, it is merely a naming convention. By analogy with electric charge, like colour charges repel each other, and unlike colour charges attract each other. So, a particle with a red colour charge (for instance) will be attracted to particles with green or blue colour charge, and also to particles with antired, antigreen, or antiblue colour charge, whereas it will be repelled by particles with red colour charge.

Each quark can have any one of the three colour charges, and each antiquark can have any one of the three anticolour charges. So, in effect, there are *three* versions of each type of quark: red up quarks, blue up quarks, and green up quarks, for instance. (Remember, this is in addition to the conventional electric charge that quarks and antiquarks also carry.)

Gluons each carry a combination of colour *and* anticolour charge in one of eight combinations. The colour–anticolour combinations of the eight different gluons may be expressed in a variety of ways; one way includes combinations such as (red-antiblue + blue-antired)/$\sqrt{2}$. Gluons all have zero electric charge.

Leptons and photons *do not* have any colour charge associated with them.

This model helps to explain many phenomena, such as why the only possible hadrons are baryons (consisting of three quarks), antibaryons (consisting of three antiquarks), and mesons (consisting of one quark and one antiquark). Each of these composite particles is colour neutral (Figure 12.2), that is to say it has a net colour charge of zero. Any baryon must contain one quark with a red colour charge, one quark with a green colour charge, and one quark with a blue colour charge. By analogy with conventional colours: red + green + blue = white, a neutral colour with a net colour charge of zero. Likewise, antibaryons must contain one antiquark with an antired colour charge, one antiquark with an antigreen colour charge, and one antiquark with an antiblue colour charge. Again, this gives a net colour charge of zero.

Similarly, the quark–antiquark pairs that constitute a meson must have the opposite colour charge to each other: red + antired = white for instance, which is a net colour charge of zero again. (Even the proposed tetraquarks and pentaquarks are colour neutral as they effectively comprise a pair of mesons or a baryon plus a meson, each of which individually are colour neutral too.) Only particles with a net colour charge of zero are allowed to exist in an independent state, and this explains why single quarks and antiquarks are not seen in isolation. The locking up of quarks inside hadrons is referred to as confinement. Gluons do not have a net colour charge

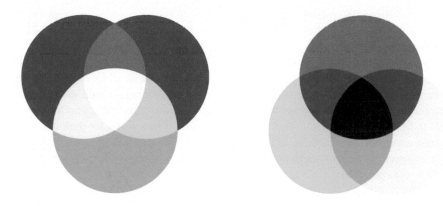

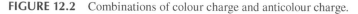

FIGURE 12.2 Combinations of colour charge and anticolour charge.

of zero either, so they too do not escape from strong interactions. Instead, gluons will decay into quark–antiquark pairs, which in turn create further hadrons.

The following points summarize the similarities and differences between the behaviour of QED and QCD:

- In QED, photons interact with electrically charged particles and their anti-particles, but photons do *not* interact directly with other photons; in QCD gluons interact with quarks and antiquarks and they *also* interact directly with other gluons, because all of these particles possess colour charge.
- Photons and leptons can escape from QED processes; gluons and quarks do *not* escape from QCD processes. Instead, they give rise to jets of hadrons (composite particles made of quarks) which do escape. The confinement of quarks inside hadrons is due to the fact that only particles with an overall zero colour charge can exist independently.
- In QED, transient electron–positron pairs cause the effective strength of electromagnetic interactions to *increase* at the short distances that are probed in high-energy experiments; in QCD the interactions between glu-ons cause the strength of strong interactions to *decrease* at higher energies.

The great success of QED and QCD was to recognize that you cannot have any one of these three differences without the other two. The self-interaction of gluons is responsible for the weakening of the strong force at higher energies and also for the confinement of quarks and gluons. Similarly, the fact that photons do not interact directly with other photons is related to the increasing strength of electromagnetism at higher energies, and accounts for the fact that electrons and photons emerge from atoms, with only a relatively small input of energy of a few eV or so.

So, when comparing the strengths of electromagnetic and strong interactions, it's necessary to take account of the energy scales that are involved, and the associated scales of distance, bearing in mind that higher energies probe shorter distances. As noted earlier, at an energy of 100 GeV, electromagnetic interactions are ten times weaker than strong interactions. But who knows what will happen at energies vastly higher than those achieved at the best laboratories on Earth? To make sense of the early Universe it is necessary to know the answer. Fortunately, the theories predict their own fates. If no new phenomenon intervenes, the strength of the QED interaction will always increase as energy increases, while that of QCD will always decrease as energy increases. It is therefore possible to estimate a rough value for the energy at which the two theories would have comparable strengths. The answer is found to be at about 10^{15} GeV. This is a million million times greater than collision energies currently attainable in high-energy particle physics laboratories. Yet cosmologists envisage early epochs of the evolution of the Universe when such collision energies were possible.

13 Weak Interactions

Weak interactions are apparent in reactions, or decays, in which some particles may disappear, while others appear. Unlike the electromagnetic or strong interactions, there is no structure that is held together by a "weak (nuclear) force." Weak interactions are vital for understanding the Universe, as they are responsible for most of the reactions in the very early Universe by which particles changed from one sort to another. They are therefore largely responsible for the overall mix of particles from which the current Universe is made up.

The most common example of a weak interaction is beta-decay occurring in an atomic nucleus. As you saw in Chapter 6, in fact there are three related processes, each of which is a different type of beta-decay: in a beta-minus decay, a neutron in the nucleus transforms into a proton with the emission of an electron and an electron antineutrino; in a beta-plus decay, a proton in the nucleus transforms into a neutron with the emission of a positron (antielectron) and an electron neutrino; and in an electron capture process, a proton in the nucleus captures an electron from the inner regions of the atom and transforms into a neutron with the emission of an electron neutrino. In each of these three processes, therefore, the nucleus involved will *change* from one type of element to another, as a result of either increasing or decreasing its proton content by one. Each of them relies on the weak interaction.

13.1 COMPARISONS BETWEEN FUNDAMENTAL INTERACTIONS

As you have seen in Chapters 11 and 12, electromagnetic interactions involve electrically charged leptons (such as the electron), quarks (all of which are electrically charged), and hadrons which are all made from quarks. Strong interactions involve only particles that possess colour charge, namely quarks and gluons, as well as composite particles made from quarks. Neutrinos are electrically neutral leptons and are involved in neither electromagnetic interactions nor strong interactions, because they possess neither electric charge nor colour charge.

The one interaction in which neutrinos do participate is the weak interaction. As the name suggests, the weak interactions of neutrinos from nuclear beta-decays are of low strength. A substantial amount of the energy released by nuclear fusion reactions inside the Sun escapes as the kinetic energy of neutrinos. The vast majority of these solar neutrinos pass through the Sun without interaction. In fact, around 70 billion neutrinos from the Sun pass through each square centimetre of your body every second, without you ever noticing. However, the probability of a neutrino interacting with matter increases with the kinetic energy of the neutrino. Beams of neutrinos with kinetic energies of the order of 100 GeV are readily obtained, as decay products, at high-energy particle accelerator laboratories. At such kinetic energies, neutrinos interact as readily with a target as electrons do. So weak interactions have roughly the same strength as electromagnetic interactions at an energy of 100 GeV, and both are about ten times weaker than strong interactions at this energy.

The key feature of the fundamental interactions presented in previous chapters is that the electromagnetic interactions of QED involve quanta called photons, while the strong interactions of QCD involve quanta called gluons. Photons interact only with particles that are electrically charged, so neutrinos are immune to them. Gluons interact only with particles that have colour charge, so all leptons are immune to them. What makes weak interactions so important is that they involve all six flavours of quark (u, d, c, s, t, and b), all three electrically charged leptons (e⁻, μ⁻, and τ⁻), all three neutral leptons (ν_e, ν_μ, and ν_τ), and all the corresponding six antiquarks and six antileptons, that you read about in Chapter 7.

As you will see below, weak interactions enable quarks to change flavour into other quarks and allow leptons to change flavour into other leptons. This is made possible by the quanta of the weak interaction, known as W and Z bosons, which are discussed next.

13.2   W AND Z BOSONS

In the same way that photons and gluons are the quanta involved in electromagnetic and strong interactions respectively, weak interactions involve other quanta – known as W bosons and Z bosons. In fact, there are two types of W boson, one with negative electric charge, the W⁻ boson, and one with positive electric charge, the W⁺ boson. The two (charged) W bosons each have identical masses of about 80 GeV/c^2 whereas the (neutral) Z^0 boson has a mass of about 90 GeV/c^2. In weak interactions, W and Z bosons interact with each other, as well as with all quarks and leptons. The Universe would be a very boring place without them.

As noted earlier, the beta-minus decay of a nucleus occurs when a neutron (with quark composition udd) turns into a proton (with quark composition uud), leading to the emission of an electron and an electron antineutrino. At most, a few MeV of energy are released in this process, corresponding to the difference in mass between the original nucleus and the resultant nucleus. At the quark level, the explanation is that a down quark (d) with a negative electric charge equal to one-third that of an electron is transformed into an up quark (u) with a positive electric charge equal to two-thirds that of a proton. A W⁻ boson is emitted with one unit of negative electric charge, thereby conserving electric charge in the process. The mass-energy of the W⁻ boson is about 80 GeV, so it cannot possibly emerge from the nucleus as there are only a few MeV of energy available. However, it can exist for a very short time, in accord with the energy–time uncertainty principle. It subsequently decays to produce an electron (e⁻) and an electron antineutrino ($\overline{\nu}_e$), setting the energy accounts straight (Figure 13.1).

As you might guess, beta-plus decay can be understood as a very similar process. In this case, a proton (uud) turns into a neutron (udd), with the emission of a positron and an electron neutrino. At the quark level, the explanation this time is that an up quark (u) with a positive electric charge equal to two-thirds that of a proton is transformed into a down quark (d) with a negative electric charge equal to one-third

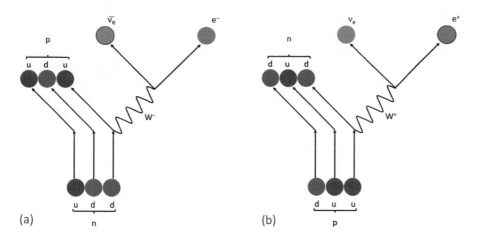

FIGURE 13.1   The processes of (a) beta-minus and (b) beta-plus decay at the level of quarks.

that of an electron. A W^+ boson is emitted with one unit of positive electric charge, thereby conserving electric charge in the process. It subsequently decays to produce a positron (e^+) and an electron neutrino (ν_e), setting the energy accounts straight again. Electron capture decay is similar, except that here, the W^+ boson liberated by the proton to neutron conversion reacts with a captured electron to produce the electron neutrino which is emitted.

In weak interactions, the total number of quarks minus the total number of anti-quarks is the same both before and after the interaction. The number of leptons is also conserved. In the examples of beta-minus and beta-plus decay, there are no leptons initially present, and after the interaction there is one lepton and one antilepton – a net result of zero again. This is the explanation for why neutrinos and antineutrinos are produced in beta-decays. If they were not, then the rule of lepton conservation would be violated. Notice also that the production of a charged lepton is always accompanied by the corresponding flavour of neutrino. In *all* weak interactions:

- Electric charge is conserved.
- The number of quarks minus the number of antiquarks is conserved.
- The number of leptons minus the number of antileptons is conserved.
- Flavour changing of either quarks or leptons is allowed, if the above three rules are obeyed.

The third type of quantum involved in weak interactions is the Z^0 boson with zero electric charge. An example of the type of reaction involving the Z^0 boson can occur in a collision between an electron (e^-) and a positron (e^+). This can create a Z^0 boson from the mass-energy of the electron–positron pair, which subsequently decays into a muon neutrino (ν_μ) and a muon antineutrino ($\bar{\nu}_\mu$). Notice that in this case there is one lepton and one antilepton both before and after the interaction.

Because the mass of a Z^0 boson is about 90 GeV/c^2, an energy of 90 GeV is needed to create one. By selecting the energy of the electron and positron beams in the Large Electron Positron (LEP) collider to be 45 GeV each, so that the total energy of 90 GeV matched that required to create Z^0 bosons, a high rate was achieved for the production of neutrino–antineutrino pairs in the process referred to above. This experiment produced an important piece of information for understanding the early Universe: there are no more types of neutrino than the three already discovered and listed in Chapter 7, i.e. ν_e, ν_μ, and ν_τ. If there were a fourth type of neutrino, the rate of electron–positron annihilation would have been higher than what was seen. As noted in Chapter 1, this was good news for cosmologists, who needed this information to calculate the rate at which nuclei were formed when the Universe was a few minutes old. Knowing that there are only three types of neutrino, cosmologists can compute the fraction of nucleons that survived as neutrons in (mainly) helium nuclei, a few minutes after the Big Bang.

13.3   THE SURVIVAL OF THE NEUTRON

Apart from hydrogen, nuclei made solely of protons cannot exist. Neutrons are necessary to make nuclei stable, so the neutron is vital to the Universe. Without it there would be only a single element, hydrogen, making chemistry extremely dull, as it would be limited to a single molecule, H_2, with no one to study it!

The rules of strong interactions allow a neutron (udd) to be made in the same manner as a proton (uud). Indeed, as you will see in Chapter 16, in the first moments of the Universe it is believed that protons and neutrons were created in *equal* numbers. Nowadays, however, the Universe contains only about one neutron for every seven protons, and almost all of those neutrons are locked up inside helium nuclei. Clearly then, at some stage, neutrons have "disappeared" from the Universe. How did this happen?

The mass-energy of a free neutron is about 1.3 MeV *larger* than that of a free proton. This energy difference exceeds the mass-energy of an electron (which is about 510 keV or 0.5 MeV) and means that free neutrons (i.e. neutrons not bound within atomic nuclei) can undergo beta-minus decay, converting into a proton plus an electron and an electron antineutrino, with a half-life of about ten minutes. This is believed to be the mechanism by which the proportion of neutrons in the Universe decreased from one in every two hadrons soon after the Big Bang, to only around one in seven today. Once neutrons are bound up into helium nuclei they no longer undergo beta-minus decay, because helium nuclei are stable.

But there is still a puzzle: if free neutrons can decay into protons, how did the neutrons form helium nuclei in time to avoid the fate of decay that affects them when they are free? It really was all down to timing. As you will see in Chapter 16, the temperature of the Universe had fallen to a value that allowed the formation of helium nuclei only a couple of minutes after the Big Bang. Because free neutrons survive for about ten minutes before decaying, there were still plenty of them around at this time, and all those that had not yet decayed into protons were rapidly incorporated into helium nuclei. But if free neutrons only survived for, say, one second, there would not have been many neutrons left to form nuclei a few minutes after the Big Bang. The vast majority of them would have long since decayed into protons. The relatively long lifetime of a free neutron is due to the fact that weak interactions (such as beta-minus decay) really are weak, and therefore occur only rarely at low energies.

As you can now appreciate, there is a vital condition for life in the Universe: weak interactions must be really weak at low energies. If they were as strong as electromagnetic interactions at low energies, beta-minus decay processes would happen much more easily and the lifetime of a free neutron would be much shorter. As a result, the vast majority of the neutrons in the Universe would have decayed before it became possible for them to become incorporated into atomic nuclei, and there would have been no elements other than hydrogen in the Universe! Yet, at high energies, such as the 100 GeV of the LEP collider, weak interactions are comparable in strength to electromagnetic interactions, and so are only ten times weaker than strong interactions, as noted earlier. You may wonder, how is this fine-tuning accomplished?

It turns out to result from the large masses of the W and Z bosons, which (as mentioned earlier) are each roughly 100 GeV/c^2. In order for any weak interaction to occur, a W or Z boson must be created. But it is difficult to produce the massive W and Z bosons when the available energy is only 1 GeV. Consequently, at an energy scale of 1 GeV, where they were first investigated, weak interactions really are weak. In contrast, at an energy scale of 100 GeV, weak interactions are not so weak. At this energy, the strength of the weak interaction is roughly the same as that of electromagnetic interactions. At this energy, W and Z bosons are easily created from the energy available. Going down a factor of 100 in energy, from 100 GeV to 1 GeV, gives a huge decrease in the rates of weak processes. At an energy scale of 1 GeV, the strength of weak interactions is a hundred million times less than it is at 100 GeV.

14 Gravitational Interactions

Last on the list of the four fundamental interactions is gravity. As you probably know, gravity was first explained by the work of Isaac Newton in the 17th century. However, there are two distinct additions that have since been made to Newton's description of this phenomenon. The first, involving relativity, was completed by Albert Einstein in 1915; the second, involving quantum physics, has only just begun.

14.1 ORBITS AND KEPLER'S LAWS

The idea of a "force of gravity" controlling the orbits of the planets didn't arise until around 350 years ago, but studies of planetary motions had been carried out long before this. The Danish astronomer Tycho Brahe working in the late 16th century made very detailed measurements of the positions of the planets as they moved through the sky, which were to prove vital in understanding their orbits. Tycho himself believed in a geocentric (Earth-centred) system in which the Moon and Sun orbited the Earth, whilst only the other planets orbited the Sun. However, it was his measurements that were used by the German mathematician and astronomer Johannes Kepler working in the following decades to come up with a correct description of how the planets (including the Earth) each orbit the Sun. This was in accord with the so-called heliocentric (Sun-centred) theory that had been suggested by the Polish mathematician and astronomer Nicolaus Copernicus almost a century earlier.

Kepler's first and second laws of planetary motion were published in 1609. The first law states that the planets each orbit the Sun in elliptical orbits, with the Sun at one of the two foci of the ellipse. In more general terms, all orbits can be described as conic sections, which are simply the curves obtained by slicing through a cone (Figure 14.1). An ellipse is the closed curve obtained by slicing through the cone at some arbitrary angle, whilst a circle is just a special case of an ellipse obtained by slicing through the cone parallel to its base. The other two types of conic section are both open curves, known as a parabola and a hyperbola. Although the planets each follow elliptical (albeit almost circular) orbits, long-period comets may orbit the Sun with parabolic or hyperbolic orbits, passing close to the Sun once, before heading off into space never to return.

An ellipse may be characterized by the lengths of its semi-major axis and semi-minor axis – namely half the distance from the centre of the ellipse to its edge, measured in the long and short directions respectively. The foci of the ellipse are two special points on its major axis, equally spaced either side of the centre, such that the sum of the distances from each focus to any point on the circumference is a constant, and equal to the length of the major axis. The eccentricity of an ellipse is the ratio of the distance between the two foci to the length of the major axis. In a circle, the two foci coincide at the centre, the semi-major axis and semi-minor axis are both equal

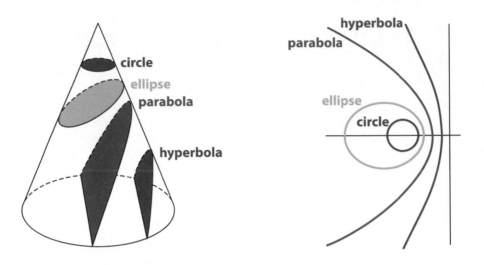

FIGURE 14.1 Orbits as conic sections.

to the radius of the circle, and the eccentricity is zero. A long, thin ellipse will have an eccentricity approaching a value of one.

Kepler's second law of planetary motion states that a line joining a planet and the Sun sweeps out equal areas in equal times. In other words, when a planet is closest to the Sun (at the end of the elliptical orbit nearest to the focus where the Sun lies) it will travel faster than when it is at the other extreme of its orbit. This law is in fact a consequence of the principle of conservation of angular momentum that was mentioned in Chapter 2.

The third law of planetary motion was only discovered by Kepler several years later and published by him in 1619. It states that the square of the orbital period of a planet is directly proportional to the cube of the semi-major axis of its orbit. Consequently, if the semi-major axis of a planetary orbit were to be four times larger than that of the Earth, its orbital period would be eight times longer (since $8^2 = 4^3$). As you know, the average distance of the Earth from the Sun is defined to be 1 astronomical unit (about 150 million kilometres) and of course the Earth takes one year to complete its orbit. According to Kepler's third law, a planet orbiting the Sun at a distance of 4 astronomical units would take eight years to do so.

14.2 NEWTON'S GRAVITY

It was the English scientist Isaac Newton who explained that Kepler's laws of planetary motion result from a universal force of gravitational attraction that acts between all objects that possess mass. The law that describes the force was discovered by Newton probably around 1666, but not published by him until about 20 years later in his *Mathematical Principles of Natural Philosophy*. It may be stated as follows:

Newton's law of gravity: two particles attract each other with a gravitational force that is proportional to the product of their masses divided by the square of their separation.

Notice the close similarity between Newton's law of gravity and Coulomb's law for electrical force. Where Coulomb's law involves the product of two electric charges, Newton's law involves the product of two masses. Also, both forces decrease with the inverse of the square of the separation between the particles. Following Coulomb's law and the relationship between brightness and luminosity, Newton's law is therefore the third example of an inverse square law that you have met.

This is where the similarity between Coulomb's law and Newton's law ends. It turns out that the gravitational force of attraction between subatomic particles, such as protons and electrons, is negligible. Gravitational forces are 43 orders of magnitude smaller (i.e. 10^{43} times smaller) than electric forces, between individual electrons. Remember that Chapter 1 talked about 45 orders of magnitude in length scales separating quasars from quarks; here there are 43 orders of magnitude separating the strength of the gravitational force from the strength of the electrical force.

Gravity is dominant only when there are large aggregates of particles, feeling no other force. The strong and weak interactions of nuclei have very short ranges, so they make no contribution to the force between, say, an apple and the Earth. But why should gravitational forces dominate in this situation instead of electric forces? It's not that the gravitational force acts over larger distances than the electric force – Coulomb's law and Newton's law imply that both gravitational and electric forces act over large distances with the *same* kind of inverse square law. The key is that electric forces can be attractive *or* repulsive because objects can possess either positive or negative electric charge, and like charges repel while unlike charges attract. By contrast, gravitational forces are *always* attractive – there is no such thing as a repulsive gravitational force. The reason for this is that mass only comes in one form – "negative mass" and "antigravity" remain in the realm of science fiction.

So, the reason that gravity dominates the interaction between an apple and the Earth is that they are both electrically neutral, to very high precision. In order for the electric force of repulsion between an apple and the Earth to be similar to the gravitational force of attraction between them, only 1 atom in every hundred billion billion would have to lose an electron. The downwards fall of the apple is due to the fact that matter is electrically neutral to a precision far better than 1 part in 10^{20}.

As a consequence of gravity, therefore, objects near to the surface of the Earth possess a property that is referred to as weight. The weight of an object is defined as the magnitude of the force of gravity acting on that object. In everyday situations, it may be expressed as the object's mass multiplied by the acceleration due to gravity at the surface of the Earth, which is almost 10 m s^{-2}.

Once again, there are similarities between how gravitational forces and electromagnetic forces behave. This expression for the gravitational force is rather similar to the expression linking electric force and electric field strength that you met in Chapter 11. The first refers to the force acting on a particle of a given mass in a region characterized by a certain acceleration due to gravity; the second refers to the force acting on a particle with a given electrical charge in a region characterized by a certain electric field strength. The concepts are clearly very similar here: gravitational forces act on objects with mass, electrical forces act on objects with electric charge.

14.3 EINSTEIN'S GRAVITY

In the decade following the publication of his special theory of relativity in 1905, Albert Einstein considered the consequences of a feature of gravity, which had been apparent since the time of Newton: you cannot use the motion of an apple, under the influence of gravity alone, to *weigh* the apple.

As you know, *mass* is a physical property that quantifies the amount of matter in a body. In contrast, *weight* is a physical property that actually quantifies the strength of the gravitational force acting on a body. Weight therefore depends on the mass of a body *and* on (say) the planet on which the body is situated. A 200 g apple and a 100 g apple have different masses and different weights; however, on Earth the two apples fall to the ground in almost the same manner. In fact, they would fall with exactly the same acceleration if it were not for the (frictional) force of air resistance. As demonstrated by the Apollo 15 astronaut David Scott on the Moon, in a vacuum, it is impossible to determine whether a hammer weighs more or less than a feather merely by watching them fall, because they do so with the same acceleration. The reason for this is that weight is directly proportional to mass. Doubling the mass of an object also doubles its weight, i.e. the gravitational force acting on it. The acceleration with which objects fall to the ground on Earth has the same value and so conveys *nothing* about either the weight or the mass.

Einstein therefore decided that it was better to think about the regions of space and time in which this motion occurs, rather than considering forces and weights. In doing so, he devised an entirely new theory of gravity. Einstein's theory of general relativity, which he published in 1915, reproduced all the old results of Newton, but without even using the idea of weight. The word spacetime was coined by Einstein's former mathematics professor, Hermann Minkowski, to reflect the fact that both space and time are inextricably linked together and interchangeable in certain circumstances. It turns out that spacetime is key to understanding Einstein's theory.

The crux of general relativity is the interaction between spacetime and the matter within it. One of its basic conclusions is that objects that possess mass change the geometry of spacetime, causing it to be "curved." The curvature of spacetime then controls the movement of material objects within it. The American physicist John Archibald Wheeler put it succinctly when he wrote, "spacetime tells matter how to move; matter tells spacetime how to curve." Gravitational effects arise as a result of the curvature of spacetime. Taking the Solar System as an example, the Sun causes spacetime to curve in its vicinity. The orbits of the Earth and the other planets are then a consequence of their movement through this curved spacetime (Figure 14.2).

An analogy that is often used is the idea of placing heavy ball bearings on a taut rubber sheet. The rubber sheet is a two-dimensional representation of the three-dimensional space of the real Universe. The large balls placed on the sheet represent

FIGURE 14.2   Curved spacetime in general relativity.

massive objects such as the Sun, which distort the space in their vicinity. The small ball bearings represent less massive objects such as planets, which move in response to the curved space through which they travel.

Does that mean that Newton's idea of massive objects producing a "force" of attraction is wrong? Well, not exactly – Newton's law of gravitation is able to predict what is observed in a wide variety of situations. But it is only an approximation to the real world and no longer provides satisfactory answers when the masses involved become very large. Similarly, while Einstein's general relativity can cope with such situations accurately, that too fails when physicists try to unite it with the ideas of quantum physics. So general relativity too may be only an approximation to some deeper, underlying truth. That idea will be revisited later in this chapter. For now though, let's take a look at some of the phenomena observed in the Universe today that can only be explained by Einstein's gravitational theory, general relativity.

14.4   TESTS OF CURVED SPACETIME

The first test of general relativity was directly related to the effects of curved spacetime and concerned the planet Mercury. Like all the planets in orbit around the Sun, Mercury moves in an elliptical path. For many years it had been known that the orientation of Mercury's orbit shifts slightly with time – an effect known as precession. In particular, the point where Mercury is closest to the Sun advances by 0.159 degrees per century. Ever since the mid-19th century it had been known that perturbations of Mercury's orbit by the other planets, as predicted by Newtonian gravity, could account for a shift of 0.147 degrees per century, but what of the remaining 0.012 degrees shift per century? As you may have guessed, general relativity was able to account for the anomalous shift in a natural way by considering the curvature of space around a massive object like the Sun.

Another consequence of the curved spacetime in the vicinity of massive objects is that light, or any other electromagnetic radiation, passing close to a massive object will have its path bent. However, saying that gravity bends the path of light is perhaps not the correct way to look at things from the point of view of general relativity. Rather, light continues to travel in a straight line, it is just that spacetime is curved near to a massive object, so (from sufficiently far away) its path appears to deviate.

The second test of general relativity that was available in the early 20th century concerned this prediction that the path of a beam of light would be "bent" as it passed close to a massive object. To an observer, watching from the Earth, it would *seem* as though the light from a distant star, passing close to the Sun, has its path bent. For a star located along a line of sight passing close to the edge of the Sun, this would have the effect of making the apparent position of the star appear different from its true position.

The problem with a test of this kind, in the early years of the 20th century, was how can you see stars that are close to the Sun? The sky is simply too bright during the daytime. The answer is that you wait until a total solar eclipse when you *can* see stars close to Sun! So, in 1919, the British astronomer Arthur Eddington sailed off to the island of Principe off the west coast of Africa and took photographs of the Sun during a total solar eclipse that was visible from there in May. When the relative positions of the stars were compared with their normal positions (as visible at night-time six months later or earlier) the deflections were found to agree well with the predictions of general relativity. The amount of shift is rather small – for a star exactly on the limb of the Sun, the angular shift is only about 0.0005 degrees! Not surprisingly, the measurements made by Eddington and others around the same time had large uncertainties in the angular measurements, typically around 30%.

The situation changed with the development of radio astronomy during the 1960s. The first important step was that radio telescopes were constructed that could determine angular positions to a precision of better than one milliarcsecond. The second important development was the discovery of the class of bright, point-like radio objects known as quasars, mentioned earlier. These are bright enough at radio wavelengths for them to be detected close to Sun without the need to wait for a total solar eclipse. A number of measurements of the deflection of radio waves from quasars as they pass close to the Sun have now been made, and they confirm the predictions of general relativity to a precision of better than 1%.

A phenomenon related to that described above became apparent in 1979 when astronomers discovered what appeared to be a double quasar. The two images had similar spectra and indeed their spectra had the same redshift as each other. When one of them became brighter, so did the other one. It was soon realized that this was not two quasars, but two images of just one quasar! The situation arises in an effect known as gravitational lensing (Figure 14.3). The idea is that a massive, but dim, galaxy lies almost directly along the line of sight to the distant quasar. Light and other electromagnetic radiation from the quasar is bent by the curvature of space in the vicinity of the intervening massive galaxy. This gives rise to two (or more)

FIGURE 14.3   Gravitational lensing. (Credit: ESO/M. Kornmesser.)

images of the lensed object. Since this first discovery many more such gravitational lensing systems have been found. A variety of patterns is possible, depending on the mass distribution in the intervening galaxy, and just how closely the distant quasar lines up behind it: quadruple patterns and arc-like structures are possible, as well as simple paired images. If the alignment is perfect, the distant object is lensed into a complete circle – known as an Einstein ring. Effects such as these simply don't happen in Newton's gravity.

Even more recently, gravitational microlensing by individual stars has also been measured. Within the Milky Way the stars are all in constant motion as they orbit around the centre of the Galaxy. From a viewpoint here on Earth, it will happen that occasionally two stars will line up with each other. When this happens, the foreground star will gravitationally lens the light from the background star, resulting in an amplification of the star's brightness that lasts for typically a few weeks as the precise stellar alignment persists. The effect had been predicted by Einstein, but he didn't think it would ever be possible to simultaneously monitor the millions of stars necessary to observe this very rare phenomenon. Nonetheless, modern survey techniques have indeed made it possible to detect gravitational microlensing, thanks to the combination of wide-field telescopes and automated analysis of the millions of stellar light curves that are measured. Surveys of the centre of the Galaxy with the Polish Optical Gravitational Lens Experiment (OGLE) have discovered many hundreds of such events.

Even more exciting is the possibility of gravitational microlensing by planets around other stars, known as exoplanets. If the foreground (lensing) star happens to have one or more planets that also line up with the background (lensed) star, then an additional lensing event will happen due to the gravity of the planet. At the time of writing, around 100 exoplanetary systems have been discovered in this way, including some systems with two planets each.

One more prediction of general relativity is the idea of gravitational redshift. This implies that when light emerges from the deep gravitational potential well around massive objects, its wavelength will be redshifted (or equivalently, the photons will lose energy). Astronomically, gravitational redshift of the light emitted by white dwarf stars was first observed from the white dwarf 40 Eridani B in 1954. This was further verified experimentally using ground-based experiments in 1959 when the gravitational redshift of the gamma-rays emitted by a radioactive source was measured over a vertical distance of 22.5 m on Earth. More recently, in 1976, this was tested using a microwave laser on a high-altitude rocket which was compared to a similar one on the ground. The experiment confirmed the existence of gravitational redshift to a precision of better than 1/100 of 1%. In fact, the global positioning system (GPS) satellites must account for gravitational redshift in their onboard timing system in order to determine their precise positions, and without accounting for the fact, they would not work at all.

14.5   GRAVITATIONAL RADIATION

A final prediction of Einstein's theory of general relativity concerns what is known as gravitational radiation. When electrically charged particles accelerate (i.e. when they change their speed or direction of motion) they emit electromagnetic radiation. In a similar way, general relativity predicts that when massive objects accelerate (due to either a change in their speed or direction of motion) they will emit gravitational radiation. Sources of gravitational waves will therefore include compact binary stars which contain a white dwarf, neutron star, or black hole orbiting around another (similar or different) compact object. The gravitational waves emitted by such systems can be thought of as ripples in the geometry of space, spreading out from their origin, stretching and squeezing space itself as they pass by. They travel at the speed of light but, in all everyday Earth-bound experiences, their effects are extremely tiny and very difficult to detect.

However, even before highly sensitive gravitational wave detectors were built, there was already excellent evidence that gravitational radiation does indeed exist, and astronomers could even point to a system where they know it is being produced right now. In 1974, US astronomers Joe Taylor and Russell Hulce discovered a remarkable binary star system: two neutron stars orbiting around each other once every 8 hours. One of the two is also a pulsar, producing regular pulses of radio waves as it spins on its axis once every 59 milliseconds. This system, which became known as the binary pulsar, has proved to be an ideal test site for general relativity. It was soon discovered that the orientation of the orbit of the two neutron stars around each other shifts with time, just as the orbit of Mercury around the Sun does. But, whereas the orbit of Mercury shifts by only 0.012 degrees per century as a result of general relativity, that of the binary pulsar shifts by a massive 4.23 degrees per year!

The large shift is a consequence of the large mass of the two stars and the rapid speeds with which they are orbiting each other. By combining this measurement with detailed observations of the variation in radio pulsations from the pulsar around the orbit, it was possible to calculate the masses of the two neutron stars. The results were 1.42 times the mass of the Sun for the pulsing neutron star, and 1.40 times the mass of the Sun for its companion neutron star. General relativity had been used to "weigh" a pair of neutron stars!

An even better test of general relativity was in store. As the system was monitored over several years, it was discovered that the 8-hour orbital period of the binary pulsar was changing very slightly: the period becomes shorter by 75 microseconds every year. This implies that the two stars are getting closer together, and that the system is losing energy. But where is the energy going? The answer is that the system is steadily giving off gravitational radiation. Calculations showed that the rate of energy loss by gravitational radiation, as predicted by general relativity, ties in exactly with the measured changes in the orbital period of the two neutron stars. Although the gravitational radiation from binary pulsars is too weak to be detected by current detectors, there is no doubt that general relativity is being seen in action in this remarkable system.

In principle, gravitational radiation may be detected directly by the effect that it has on objects here on Earth. Current detectors, including the American *LIGO* detectors and the European *Virgo* detector, consist of two extremely long beams of laser light (typically a few kilometres in length) orientated at right angles to each other (Figure 14.4). When a gravitational wave passes through the device it will

FIGURE 14.4   The *Virgo* gravitational wave detector.

cause very slight changes in the length of one beam relative to the other. This change in length may be measured by means of the variation produced in an interference pattern formed by combining the two beams. The big problem is that the distortions produced by the passage of the gravitational wave are extremely tiny. The first detectors to be built are capable of measuring relative length changes of 1 part in 10^{21}. To put this in perspective, if the laser beams in the detectors stretched from the Sun to the nearest star, the equivalent distortion in length would be about the same as the thickness of a human hair!

In February 2016, almost exactly 100 years after Einstein's prediction, the first direct detection of gravitational waves was announced by the team running the Laser Interferometer Gravitational Observatory (*LIGO*). On 14 September 2015 at 09:51UT, *LIGO* detected the characteristic "chirp" of gravitational waves resulting from the in-spiral and coalescence of a binary black hole system. Analysis of the signal showed that it resulted from a pair of black holes, with masses about 29 and 36 times that of the Sun. During the spiral together and subsequent merger, an amount of matter equal to about 3 times the mass of the Sun was converted into the energy of gravitational waves in a fraction of a second, leaving behind a single black hole with a mass of 62 times that of the Sun.

In the years since this first detection, other black hole–black hole mergers have been detected and, as you read at the end of Chapter 8, mergers of neutron stars have now been detected too, heralding a new era of multi-messenger astrophysics as such events also release vast amounts of electromagnetic radiation.

The next generation of gravitational wave detectors is even more ambitious. A project called the Laser Interferometer Space Antenna (*LISA*) proposes to fly three spacecraft in a triangular formation, each separated from its neighbours by 5 million kilometres, and following the Earth in its orbit around the Sun, about 50 million kilometres behind us. The instruments carried by each spacecraft will monitor continuously the separation between them to detect tiny distortions of space caused by passing gravitational waves.

14.6   QUANTUM GRAVITY

The final remarkable feature of gravity is that no one has yet worked out a convincing way of combining quantum physics with general relativity. It's more than a century since Einstein published his theory of gravity, but a relativistic quantum theory of gravity (usually referred to simply as a theory of quantum gravity) still does not exist. There are two reasons why such a theory has not (yet) been formulated.

First, there is essentially no experimental data on the way in which gravity and quantum physics interact – simply because gravity is so weak at the level of individual particles. The second problem is a conceptual one. Earlier it was noted that the quantum physics of atoms relies on Heisenberg's uncertainty principle and so involves uncertain velocities and positions for the electrons. In a quantum theory of gravity, somehow the notions of space and time would also have to become uncertain. In non-gravitational quantum physics you cannot be sure exactly what you will measure, here and now, or there and then. In a quantum theory of gravity, you would be unsure of what here and now, or there and then, even mean!

Despite these difficulties, it is believed that quanta of gravitational energy exist. They are known as gravitons and have zero electric charge and zero mass. It is predicted that gravitons interact with everything: not just material bodies with mass, but also photons and gluons (which have no mass) and neutrinos (whose masses are tiny, but poorly known). Gravitons are also predicted to interact with other gravitons, in a similar way that gluons can interact with other gluons, and the W and Z bosons can interact with other W and Z bosons. That makes photons the only one of the quanta that mediate the four fundamental interactions that do not interact with themselves.

The history of science is full of examples where new experiments led to new theories to explain the results, as happened in the early days of quantum physics, and other examples where new theories led to new experiments to test them, as in the case of the work carried out by Isaac Newton, James Clerk Maxwell, and Albert Einstein. However, it is rare for progress to be limited in both experiments and theories. In the recent scientific literature there are plenty of speculative ideas about what a quantum theory of gravity might involve, one of which will be mentioned in Chapter 15. The difficulty of imagining what a quantum theory of "fuzzy" space and time might be will set a limit to how far back in time the history of the Universe can be traced.

15 Unified Theories

The full set of four fundamental interactions were outlined in Chapters 11–14, so we can now summarize comparisons between them by looking in turn at their quanta, ranges, theories, participants, and strengths.

In terms of their quanta, strong interactions involve gluons; electromagnetic interactions involve photons; weak interactions involve W^+, W^-, and Z^0 bosons; and gravitational interactions involve gravitons, though experimental evidence for these is still lacking.

Considering the range over which they operate, electromagnetic and gravitational interactions both have a very large range (in principle, infinite), but the strength of both forces decreases with the inverse square of distance. Electromagnetic energy is radiated by electric charges that accelerate, and this energy propagates through space as an electromagnetic wave. Similarly, gravitational energy is radiated by massive objects that accelerate, and this energy propagates through space as a gravitational wave. Both the strong and weak interactions have a very small range, comparable to the size of individual nuclei.

When we consider the theories behind these interactions, we note that strong, electromagnetic, and weak interactions are each well described by relativistic quantum theories. The first to be developed was quantum electrodynamics (QED), for electromagnetic interactions. This involved combining quantum physics with special relativity and the inclusion of phenomena such as transient electron–positron pairs. Quantum chromodynamics (QCD) describes the strong interaction in a similar way, with the key difference that the quanta which are exchanged, called gluons, interact with themselves, as well as with quarks. This results in the permanent confinement of quarks and gluons within hadrons and means that quarks and gluons have never been seen in isolation. The theory of the weak interactions also involves quanta that may interact with each other; they are known as W and Z bosons. Gravity awaits unification with quantum physics. This makes it hard to get good data on the interplay of gravity and quantum physics. It is also hard to get good ideas about what such a theory would involve, since it must somehow incorporate a "fuzziness" of space and time.

Turning now to the participants in each interaction, only quarks take part in strong interactions, whereas quarks and charged leptons (i.e. not neutrinos) take part in electromagnetic interactions. However, all quarks and all leptons take part in both gravitational and weak interactions.

Finally, considering the strengths of each interaction, we see that the strong, electromagnetic, and weak interactions have strengths that differ only by a factor of ten at energies of 100 GeV. As expected from its name, the strong interaction is stronger than the other two. The weak interaction is very weak at low energies, where there is a big price to pay for exchanging the W and Z bosons whose mass energies are around 100 GeV each. However, at energies of 100 GeV, the weak interaction becomes comparable in strength with the electromagnetic interaction. Again, gravity is very different, due to its extreme weakness at the level of individual particles.

According to what we currently know, there are *no more than* these four fundamental interactions of all matter and radiation. Physicists and cosmologists are especially interested in trying to find out whether these four interactions really are so distinct, or whether they might be different facets of some more basic unity of nature.

The idea of unifying descriptions of force is not new: you saw in Chapter 11 that Maxwell was able to unify the (apparently disparate) phenomena of electricity, magnetism, and light. In the late 20th century there was a lot of activity in investigating a scheme for the unification, at high energies, of weak interactions with electromagnetism. This is known as *electroweak unification* and is the subject of Section 15.1 (Figure 15.1). The consequences are being tested by current particle accelerators, and some answers are now already known.

As a result of this success, some theoretical physicists have suggested a unification of the electroweak theory with QCD: the theory of the strong interaction. This is the subject of Section 15.2, on the so-called *grand unification*. For this theory, testable predictions are harder to come by. Perhaps the Universe itself is the best laboratory?

Finally, so that gravity is not left out of the picture for unified quantum theories, attempts have also been made at what is called *superunification*, which is mentioned briefly in Section 15.3. These ideas are highly speculative, but they may give us clues about what happened in the very early Universe.

Sections 15.1–15.3 are designed to give you a flavour of current research. Unlike previous parts of the book, they contain ideas which are the subject of intense debate amongst scientists. Also note that the word "unification" is here used in an *active* sense. This chapter is not about armchair discussions of whether things are different

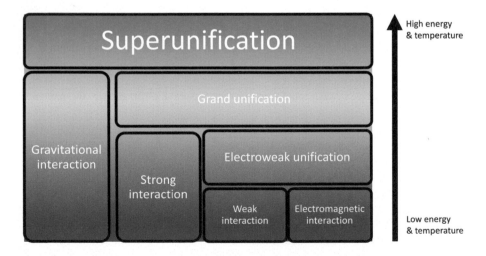

FIGURE 15.1   Unifications of fundamental interactions.

or similar. It involves real questions about the behaviour of quarks and leptons, at very high energies, and so it will feature in the account of the history of the Universe given in Chapter 16.

The following analogy, concerning the halogen elements, may help. At room temperature: iodine is a solid, bromine is a liquid, and chlorine is a gas. The substances are clearly different, in crucial respects. Whatever we say or think, they are not the same thing. But if you raise the temperature, so that iodine melts, you might find that liquid iodine and liquid bromine are quite similar in some ways, which you could not appreciate at room temperature. That might lead you to develop a "unified theory of iodine and bromine." Increasing the temperature further, so that both iodine and bromine evaporate and turn into gases, you might find strong similarities between iodine, bromine, and chlorine as gases. Then you might come up with a "unified theory of iodine, bromine and chlorine." In fact, as noted, all three elements are halogens and do indeed have some similar properties. It is a *little* bit like that with the weak, electromagnetic, and strong interactions respectively.

The significance for cosmology is that there was, as you have seen in Chapter 10, an early hot epoch of the Universe when weak and electromagnetic interactions may have been more unified than they now appear. There was probably an even earlier, hotter, epoch when electroweak interactions and strong interactions may have been unified. If so, radically new processes that turn quarks into leptons, and vice versa, may have been in operation, leading to a possible explanation of features of the currently observable, far cooler Universe.

15.1   ELECTROWEAK UNIFICATION

You know, from Chapter 13, that the large masses of the W and Z bosons are responsible for the long lifetime of free neutrons, and the very feeble interactions of low-energy neutrinos from nuclear beta-decay. The similarity in strength of electromagnetic and weak interactions only becomes apparent when comparing the interactions of electrons and neutrinos with kinetic energies of around 100 GeV, or greater; well below that energy, there is a huge difference between their behaviours.

A mechanism to explain the high-energy electroweak unification, and the lower-energy difference, was proposed in the 1960s by several theoretical physicists, including Sheldon Glashow, Steven Weinberg, and Abdus Salam. The proposal relies on the existence of a new particle, called the Higgs boson, which is named after another of the scientists involved, British physicist Peter Higgs. Around 2010, the construction of a new particle accelerator, the Large Hadron Collider (LHC) at CERN near Geneva, was completed. Then, in 2012, the results of the first experiments were announced, suggesting that the Higgs boson does indeed exist, with a mass energy of about 125 GeV.

Just why the Higgs boson should exist is a complicated tale, but the following discussion will give you the general idea. A unified electroweak theory must be able to account for all three quanta involved in the weak interaction (W^+, W^-, and Z^0 bosons) as well as the photon that is involved in the electromagnetic interaction. The problem is that the W and Z bosons have mass and interact with each other, whereas photons are massless and do not interact with other photons. Photons cannot have mass, otherwise there would be no such thing as Coulomb's law. Massive quanta – such as the W and Z bosons – cannot produce an inverse square law of force; their effects decrease much faster with distance. The problem is to develop a theory that explains the existence of four quanta, three of which are different from the other one.

According to the current theory of electroweak unification, there are four so-called "Higgs fields," one corresponding to each of the W^+, W^-, and Z^0 bosons and the photon. Three of these fields "give mass" to the W and Z bosons. The fourth field does not give mass to the photon but should be detectable as a true particle – the Higgs boson itself. If the Higgs boson does indeed have a mass energy around 125 GeV, then well above this energy range, the electromagnetic and weak interactions will appear truly unified and merely be different aspects of a single electroweak interaction.

As you can see, the story of the Higgs boson is a complicated one, but it is good science: to get a satisfactory explanation of effects at energies currently observed, theoretical physicists have made predictions of what should be seen at a higher energy. The issue of electroweak unification is also important for cosmology: how can we hope to follow the story of the Universe back to times when the energies were enormously higher than 1000 GeV, if there is a problem in the region between 100 GeV and 1000 GeV?

15.2   GRAND UNIFICATION

Even though electroweak unification is now reasonably well confirmed, this still leaves the strong and gravitational interactions out of the picture. Further unification of the forces of nature is an obvious theoretical challenge.

As already mentioned, the strength of electromagnetic interactions (according to QED) *increases*, rather slowly, with the interaction energy involved in the process. The corresponding measure of the strength of the strong interactions (according to QCD) *decreases* with the interaction energy, again rather slowly. This raises two questions: what is the energy scale at which they might become equal? and are there any new processes that might occur at this energy scale? Such grand unification of strong and electroweak interactions would only leave gravity un-unified with the rest.

The proposed energy scale for grand unification is about 10^{12} times higher than that which can be achieved with current particle accelerators. As noted in Section 12.2, it is of the order of 10^{15} GeV, as compared with the energies of the order of 1000 GeV available at the Large Hadron Collider. Such very high energies were probably involved in the early Universe, but they will not be achieved, by human means, on Earth, for the foreseeable future. Above this energy it is predicted that there is a single interaction between particles that is characterized by a single strength.

The answer to the second question concerning possible new processes is rather intriguing. The expectation is that there are quite new interactions, involving bosons with mass energies of around 10^{15} GeV. These hypothetical particles have been named X bosons and Y bosons, because nothing is known about them, directly, from experiment.

The prediction is that these new interactions allow quarks to change into leptons, and they also allow matter to change into antimatter, and vice versa in each case. The hypothetical X boson has an electric charge of +4/3 that of a proton, whilst the hypothetical Y boson has an electric charge of +1/3 that of a proton. As usual, their antiparticles ($\bar{X}$ and $\bar{Y}$) carry the opposite charge. They interact as follows:

$$\text{quark} + \text{quark} \longleftrightarrow X \text{ or } Y \longleftrightarrow \text{antiquark} + \text{antilepton}$$

$$\text{antiquark} + \text{antiquark} \longleftrightarrow \bar{X} \text{ or } \bar{Y} \longleftrightarrow \text{quark} + \text{lepton}$$

Notice that an equal number of X and Y bosons (and their antiparticles) would produce three quarks and antiquarks for every one lepton and antilepton. This will be important in understanding how the early Universe developed.

The reason that such processes have not yet been seen working at energies far below 10^{15} GeV is similar to the reason why neutron decay is a (relatively) slow process. Remember that weak interactions, responsible for neutron decay, involve W and Z bosons and are indeed very weak at energies far smaller than the 100 GeV that corresponds to the mass of the W and Z bosons. Processes involving X and Y bosons would also be very weak at energies far smaller than the 10^{15} GeV that corresponds to the mass of an X or Y boson and would be incredibly slow at the energies that are currently observable (below 1000 GeV). The behaviour of the hot early Universe would depend crucially on the processes mediated by X and Y bosons, but what is observed at lower energies hardly depends on them at all.

However, the new processes *might* show up, very rarely, at lower (i.e. achievable) energies. For example, one effect of such new processes may be that protons are not stable, but instead they eventually decay. The typical proton lifetime predicted by the Grand Unified Theory is around 10^{33} years; this is immensely longer than the age of the Universe, which is (only!) around 10^{10} years. In order to actually detect proton decay, the intrinsically random nature of all subatomic processes helps: starting with 10^{33} protons (equivalent to a mass of more than 1000 tonnes) and waiting for a few years, it might be possible to observe a few decays (if only we could spot them).

Experiments approaching this sensitivity are currently in progress and might yield results within the lifetime of this book. However, even if evidence in support of grand unification emerges from searches for rare processes, such as proton decay, or (as sometimes happens in science) from something unexpected, one step will remain in the ambitious attempt to construct a coherent "theory of everything": the construction of a theory of quantum gravity.

15.3 SUPERUNIFICATION: STRINGS AND BRANES

As already discussed, in Chapter 14, Albert Einstein transformed Isaac Newton's theory of gravity as a force into a reinterpretation of space and time. His theory of general relativity involved no quantum physics, though it did greatly modify ideas about space and time. Two of the obstacles to combining it with quantum physics have been mentioned: the difficulty of introducing ideas of uncertainty into the discussion of the properties of space and time; and the lack of laboratory data that it can be tested against.

A current theory that incorporates quantum gravity involves the descriptions of particles as strings, rather than points. Up to now, the various particles (e.g. leptons, quarks, and photons) have been described as "points" that can move through space as time progresses. Each particle can then be characterized by its "position" in three-dimensional space plus one dimension of time. However, you have seen that particles have other characteristics too, such as electric charge, colour charge, mass, and spin. All of these extra characteristics have to be included to reach a complete description of the particles.

In string theory, particles are replaced by strings that can be open (meaning they have two ends) or closed (meaning they form a loop like an elastic band). Importantly, strings can vibrate, and different modes of vibration (like different notes on a violin string) correspond to particles with different masses or charges or spins. Consequently, each type of particle may be represented by a different mode of vibration. One mode of vibration makes the string look like a photon, another makes it look like a muon, yet another mode makes it look like a down quark, and so on. Crucially, there is even a mode of vibration that corresponds to a graviton.

A remarkable prediction of string theory is that spacetime actually has ten dimensions. You probably think – how absurd! After all, you only experience three dimensions of space plus one of time. Where are the other six dimensions predicted by the theory? The explanation is that the other six dimensions are curled up very small so that they are not noticeable in normal experience. Small in this context means around 10^{-35} m, which is about 10^{21} times smaller than an atomic nucleus, so it is no wonder that they are not noticeable. The reason why these extra dimensions are needed is that properties such as mass, electric charge, colour charge, and spin then correspond to movement, or vibration, of the string within these compact dimensions, just as everyday behaviour such as speed of motion corresponds to movement in the regular three dimensions of space plus one of time.

This is all well and good (even if a little bizarre), but physicists soon discovered not one, but *five* versions of string theory that are consistent and make sense. Some of these involved open strings, some involved closed strings, and some involved both types. If string theory really is the route to a theory of everything, why should there be (at least) five different varieties?

A possible answer to this problem came with the discovery that the five string theories are actually different aspects of a single underlying theory, now called M-theory. M-theory contains much more than just strings though. It contains other multidimensional objects called branes. The word comes from "membrane," which refers to any two-dimensional structure, like a rubber sheet for instance. In M-theory, however, branes can exist in up to *11* dimensions. At present, ideas such as M-theory are being investigated by some of the smartest scientists around, but whether they lead to testable predictions remains to be seen. One hope is that the difficulty of achieving superunification, of all four interactions, results from there being only one way of doing it, which has not yet been found.

The typical energy scale at which superunification might occur is known as the Planck energy, and it has a value of around 10^{19} GeV (i.e. 10^{28} eV). It is at this scale of energy (10^{16} times higher than the energies that have been studied on Earth) that the standard model for the history of the Universe will begin, in the next chapter. Ultimately, we might hope to derive the entire evolution of the Universe, from the Planck scale of superunification, at 10^{28} eV, through the X and Y boson scale of grand unification, around 10^{24} eV, down to the Higgs boson scale of electroweak unification, now under intensive experimental investigation at 10^{12} eV on planet Earth, and from there through the well-charted territory of Chapters 11–13, right down to the 2 eV photons of visible light that carry this message to you. Only time will tell whether or not an understanding of all 28 orders of magnitude lies within the grasp of the human intellect.

Part IV

The Universe through Time

In this final part of the book, the strands from the previous chapters are woven together to tell the history of the evolving Universe from the instant of the Big Bang to the present day and onwards to the far distant future. The final question posed in Chapter 1 will be answered – how does the Universe change with time?

16 The History of the Universe

This chapter attempts to provide a detailed description for the history of the Universe. The ideas presented here are the best explanation currently available for the reasons why the Universe displays the behaviour that is currently observed. In the following sections you will be reminded about all of the fundamental particles, nuclei, and atoms that you read about in Part I, as well as the origin of the large-scale behaviour that was presented in Part II. You will also see how the various interactions discussed in Part III have each played a role in determining the present-day structure and contents of the Universe.

16.1 TIME, SPACE, TEMPERATURE, AND ENERGY

The conventional view of the Universe is that, at the very instant of the Big Bang, the Universe came into being. There was no "before" this instant because the Big Bang marked the creation of time. No location for this event can be specified because the Big Bang marked the creation of space. All that can be discussed are times after the Big Bang, and things that happen in the space created as a result of it. This is a difficult concept to visualize; but please stick with it and examine the consequences that follow.

As you saw earlier, Chapter 9 concluded that the separations between distant objects are continually *increasing* with time, while Chapter 10 concluded that the temperature of the Universe is continually *decreasing* with time (Figure 16.1). This implies that the early Universe was much *denser* and *hotter* than it is today. The thread running through this chapter is therefore one in which space is expanding, and in which the temperature is falling, throughout the Universe. In the early part of its history, every time the Universe increased in age by a factor of a hundred, it also cooled by a factor of ten and distances within the Universe increased by a factor of ten. A consequence of the cooling and expansion is that the mean energy per particle (i.e. the energy available for any reaction to occur) is continually reduced. This has important implications for the ways in which the four fundamental interactions manifest themselves at different epochs.

In Chapters 11–14, you saw that the four fundamental interactions have very different strengths and act on different types of particle. Then, in Chapter 15, you saw the clue to their unification, namely that the strengths of the interactions vary with the energy of their environment. Furthermore, at very high energies, particles can transform into different types – quarks into leptons for instance. So the fact that only quarks feel the strong interaction, while leptons do not, is irrelevant at very high energies. At higher and higher energies, first the electromagnetic and weak interactions become unified as energies reach around 1000 GeV. Then the strong interaction

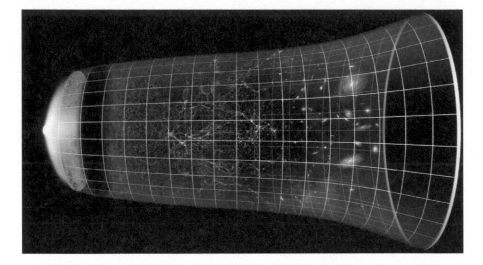

FIGURE 16.1   The evolution of the Universe. (Credit: ESO/M. Kornmesser.)

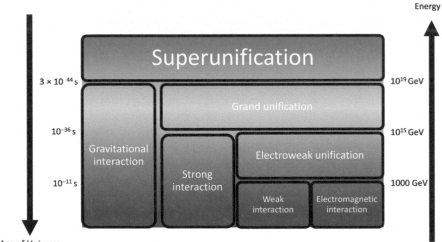

FIGURE 16.2   Fundamental interactions in the evolving Universe.

becomes unified with the electroweak interaction at an energy of around 10^{15} GeV. Finally, at the very highest energies of at least 10^{19} GeV, gravity too may become unified with all the other interactions (Figure 16.2).

At the very earliest times, the Universe was extremely hot, the mean energy available per particle was extremely high, and so the unification of interactions discussed in Chapter 15 would have occurred naturally. As the Universe has cooled, the available energy has fallen, and the interactions have in turn become distinct until the current situation is reached in which four different interactions are observed.

The rest of Chapter 16 will necessarily refer to incredibly short intervals of time after the Big Bang. Many important processes took place within the first one second after the Big Bang, when the energy available for processes in the Universe was extremely high. In fact, most of the important processes were completed by the time the Universe was less than about 15 minutes old!

16.2   THE VERY EARLY UNIVERSE

In this section, the focus is on times before the Universe was 10^{-36} s old. At this epoch, the temperature was greater than 10^{28} K and the typical energy per particle available was more than 3×10^{15} GeV.

At these very earliest times in the history of the Universe, it is presumed that a superunification of the four interactions existed. As you have seen, no reliable theory for this is yet available, so *nothing* can be said about the contents or behaviour of the Universe in its earliest moments. It may even be that the concept of "time" itself had no meaning until the Universe had cooled below a certain threshold.

The first stop on the tour where anything can be said is at about 3×10^{-44} s after the Big Bang – an epoch known as the Planck time. By this time, the mean energy per particle in the Universe had fallen to around 10^{19} GeV (the Planck energy that you met earlier). This is the energy at which the gravitational force on an individual particle has roughly the same strength as its other interactions. An idea of the typical size scale of the Universe can be gained by thinking about how far a photon of light could have travelled during this period. By the time the Universe was 3×10^{-44} s old, a beam of light travelling at 3×10^8 m s^{-1} could have travelled a distance of only about 10^{-35} m. This tiny dimension is referred to as the Planck length; it is as many times smaller than a nucleus as a nucleus is smaller than the Earth. As noted previously, this is the sort of scale on which the hidden dimensions involved in M-theory are supposed to be curled up very small.

At or around the Planck time, it is supposed that gravitational interactions became distinct from a grand unified interaction that included the three effects seen today as the electromagnetic, strong, and weak interactions. In order to describe the gravitational interactions at these times, a theory of quantum gravity is required. However, as you saw earlier, no such theory is yet available.

The temperature, and therefore the mean energy per particle, was far higher at this time than can be recreated in particle accelerators here on Earth. Cosmologists and particle physicists can therefore only speculate on what might have occurred in the very early Universe. The best guess is that pairs of matter and antimatter particles of all types were spontaneously created out of pure energy, which can be thought of as a "sea" of photons filling the entire Universe. With equal spontaneity, pairs of matter and antimatter particles also combined with each other again to produce photons. As you saw earlier, the overall process of pair creation (left to right) and annihilation (right to left) can be represented as:

$$\text{photons} \longleftrightarrow \text{particle} + \text{antiparticle}$$

At the temperatures existing in the Universe today, reactions such as this proceed preferentially from right to left. However, at the temperatures applying in the early Universe, the reactions proceeded in both directions at the same rate, for all types of particle. A stable situation was reached in which the rates of pair creation and annihilation exactly balanced, and equal amounts of matter/antimatter and radiation were maintained.

As well as the familiar quarks and leptons, if the Grand Unified Theory discussed in Chapter 15 is correct, then this is when the particles known as X and Y bosons would also have been in evidence. These particles are the quanta of the grand unified interaction and are suggested as a means of *converting* between quarks and leptons, or between matter and antimatter.

The next notable time is at about 10^{-36} s after the Big Bang when the Universe had a temperature of about 10^{28} K. This temperature marks the energy at which the strong interactions became distinct from the electroweak interactions.

It should be emphasized that there is some disagreement and uncertainty about the exact processes that occurred at this extremely early period in the history of the Universe, but the story outlined above is the best guess at what may have actually occurred. Before proceeding with the trip through time, it is necessary to pause for a moment to examine a quite remarkable event that seems to have happened just after the strong and electroweak interactions became distinct. The event has profound consequences for the nature of the Universe today.

16.3   INFLATION

We next consider a period of the Universe's history that lasted from an age of 10^{-36} s to 10^{-32} s. During this time period, both the temperature and energy per particle available were rapidly changing.

When talking about the Universe, there is an important distinction that the present discussion has, up until now, largely ignored. First, there is the entire Universe and this may be infinite in size. By implication, therefore, it makes no sense to put a value on the "size" of the entire Universe. But there is also what may be called the *observable* Universe, which is that part of the Universe that it is theoretically *possible* to be observed from Earth. A value for the size of this finite region *can* be calculated, for the following reason.

The speed of light is a cosmic speed limit – nothing can travel any faster. So, the only part of the Universe that is now observable is that fraction of it from which light has had time to reach here since the Universe began, about 14 billion years ago. You might naturally expect that the radius of the currently observable Universe is therefore equal to the maximum distance that light can have travelled since the Universe began. The calculation of this distance is complicated by the fact that the Universe has continued expanding during the time that light has travelled from distant objects (and also due to the fact that the expansion rate has slowed down and speeded up at various times, as you'll read later), but nonetheless there *is* a finite size for the observable Universe.

When trying to understand the large-scale structure of the Universe that is observed today, one of the most intriguing problems is that the Universe is so uniform. The results from the COBE and WMAP satellites showed that one part of the Universe has exactly the same temperature, to a precision of better than 1 part in 10,000, as any other part of the Universe. The expansion rate of the Universe in one direction is observed to be exactly the same as that in any other direction too. In other words, the observable Universe today is seen to be incredibly *uniform*. At 10^{-36} s after the Big Bang, when things were far closer together than they are now, the physical conditions across the Universe must therefore have been identical to an extremely high level of precision. Yet, according to conventional physics, there has not been time for these regions of space to ever "communicate" with one another – no light signals or any other form of energy could travel from one to the other and smooth out any irregularities.

In 1981, the American physicist Alan Guth suggested that, in the early history of the Universe at times between about 10^{-36} s and 10^{-32} s after the Big Bang, the Universe underwent a period of extremely rapid expansion, known as inflation. During this time, distances in the Universe expanded by an extraordinary factor – something like 10^{50} has been suggested, although this could be a vast underestimate!

It is believed that inflation may be caused by the way in which the strong and electroweak interactions became distinct. It has been likened to a phase transition, such as you get when water freezes to become ice. In that process, energy in the form of latent heat is released, and an analogous process may have occurred in the Universe during inflation. The exact mechanism by which inflation occurred is not important here, but there are many consequences of this theory. The most important for the present discussion is that the region that was destined to expand to become the currently observable Universe originated in an extremely tiny region of the pre-inflated Universe. This tiny region was far smaller than the distance a light signal could have travelled by that time and so any smoothing processes could have operated throughout the space that now constitutes the observable Universe. The problem of the uniformity of the microwave background and the uniform measured expansion then goes away.

Non-uniformities may still be out there, but they are far beyond the limits of the observable Universe – and always will be, according to current theories. Because there is no way to ever "see" beyond this barrier, there can be no knowledge whatsoever of events that occurred *before* inflation, because any information about such events is washed out by the rapid increase in scale. Inflation serves to hide from view any event, process, or structure that was present in the Universe at the very earliest times.

If you are thinking that the inflation theory contains some rather bizarre ideas – you're right, it does – but it is the most promising theory that currently exists for one of the earliest phases in the history of the Universe. No more will be said about it here, but now the story will be picked up again after the Universe has completed its cosmic hiccup. The strong and electroweak interactions have now become distinct and the X and Y bosons have therefore disappeared.

As the matter and antimatter X and Y bosons decayed, they produced more quarks, antiquarks, leptons, and antileptons – so adding to the raw materials from which the material contents of the Universe were later built. All six flavours of quark (u, d, c, s, t, and b) and all six flavours of lepton (e^-, μ^-, τ^-, ν_e, ν_μ, and ν_τ) were produced at this time, along with their antiparticles. It is important to note that matter and antimatter X and Y bosons decayed into *either* matter *or* antimatter particles. This will be important later on in the story.

16.4   THE QUARK–LEPTON ERA

During the time interval 10^{-32} s to 10^{-11} s, that is to say for the 10^{-11} seconds or so after inflation, nothing new happened in the Universe. It merely carried on expanding and cooling, but no new processes took place. The desert (as it is known) – came to an end when the Universe cooled to a temperature of about 3×10^{15} K. At this point, the mean energy per particle was around 1000 GeV and the electromagnetic and weak interactions became distinct. As you saw earlier, the energies corresponding to this transition are becoming attainable in experiments here on Earth. So, it could be argued that all particle reactions that models propose after the first 10^{-11} s of the history of the Universe are now *directly* testable in Earth-based laboratories.

Therefore, the next epoch to consider is known as the quark–lepton era, when the age of the Universe was between 10^{-11} s and 10^{-5} s. During this time period, the temperature fell from about 3×10^{15} K to 3×10^{12} K and the energy per particle dropped from 1000 GeV to 1 GeV.

By 10^{-11} s after the Big Bang, the X and Y bosons had long since decayed, but the temperature of the Universe was still too high for the familiar baryons (protons and neutrons) to be stable. The Universe contained all types of leptons, quarks, antileptons, and antiquarks as well as photons. In fact, there would have been approximately equal numbers of particles and antiparticles at this time – but note that word *approximately* – the implications of this will be considered in a moment. There would also have been equal amounts of radiation (photons) and matter/antimatter (particles or antiparticles).

The next stage of the story is to consider how and when the original mixture of all types of quarks and leptons, that were present when the Universe was 10^{-11} s old, gave rise to the Universe today, which seems to be dominated by protons, neutrons, and electrons. In the early Universe, when the mean energy per particle was greater than the mass energy of a given particle plus antiparticle, those particles and antiparticles existed in abundance, and survived in equilibrium with radiation. When the mean energy per particle dropped below this value, annihilations became more likely than pair creations, and so the number of particles and antiparticles of a given type declined.

Massive quarks and leptons also decay into less massive ones, and these decays became more likely as the available energy fell. Broadly speaking, when the temperature of the Universe fell below that at which the mean energy per particle was similar to the mass energy of the particles concerned, the particles decayed into other less massive particles.

So, by the time the Universe had cooled to a temperature of 3×10^{12} K, equivalent to a mean energy per particle of about 1 GeV, when the Universe was 10^{-5} s old, several important changes had taken place. All particles with a mass energy greater than 1 GeV had disappeared: first, many of the taus and antitaus, muons and antimuons had decayed into their less massive lepton counterparts, namely electrons and positrons. The only leptons that remained in the Universe in any significant number were therefore electrons and neutrinos (with their antiparticles in approximately equal numbers). Similarly, the massive second- and third-generation quarks (strange, charm, top, and bottom) had mostly decayed into their less massive first-generation counterparts (up and down), via a range of weak interactions involving W bosons. In each case, quarks changed flavour with the emission of a lepton–antilepton pair.

All types of quarks and antiquarks also underwent mutual annihilations – with a particularly crucial result. In discussing the relative numbers of particles and antiparticles earlier, the phrase *approximately* equal was used deliberately. If the Universe had contained *exactly* equal numbers of quarks and antiquarks, then these would have all annihilated each other, leaving a universe that contained no baryons – so no protons and neutrons – no atoms and molecules – no galaxies, stars, planets, or people. Clearly that is *not* what is observed in the Universe!

In fact, the Universe now seems to consist almost entirely of matter (rather than antimatter) in the form of protons, neutrons, electrons, and electron neutrinos, plus photons. And there are believed to be roughly ten billion photons for every baryon (proton or neutron) in the Universe today. This implies that, just before the quark–antiquark annihilations took place, for every ten billion antimatter quarks there must have been *just over* ten billion matter quarks. Running the Universe forward from this point, for every ten billion quarks and ten billion antiquarks that annihilated each other producing photons, a few quarks were left over to build baryons in order to make galaxies, stars, planets, and people.

Why did the Universe produce this slight imbalance of matter over antimatter? Maybe it was just "built-in" from the start, like any other constant of nature? This is rather unsatisfactory to many cosmologists and particle physicists who prefer to believe that the imbalance arose *after* the Universe had got started. It has been suggested that the decays of X and Y bosons into quarks and leptons *may* slightly favour the production of matter particles over antimatter particles. As mentioned earlier, a matter or antimatter X or Y boson can decay into *either* matter particles or antimatter particles. So, if there is an imbalance in the rates, starting with equal numbers of matter and antimatter X and Y bosons *will not* lead to the production of equal numbers of matter and antimatter quarks and leptons. Such matter–antimatter asymmetry has already been observed with experiments on Earth that measure the decay of particles called K mesons. (K mesons, or kaons, contain a strange quark, or antiquark, and an up or down quark, or antiquark; they have mass energies around 500 MeV.) Of the two possible routes for K meson decay, one is favoured over the other by 7 parts in 1000. Perhaps something similar, of the order a few parts in ten billion, occurs with X and Y boson decays? The answer to this question is not yet known – but it is a rather important one, because without it no-one would be here to discuss the matter! It is a rather humbling thought that the existence of the entire matter content of the Universe may be the result of an imbalance in the rates of two decay reactions by a few parts in ten billion.

16.5   THE HADRON ERA

When the temperature of the Universe reached 3×10^{12} K, at about 10^{-5} s after the Big Bang, stable baryons (protons and neutrons) began to form from the up and down quarks that remained after the annihilation of matter and antimatter. Consequently, the next epoch in the Universe's history may be referred to as the hadron era, when the age of the Universe was between 10^{-5} s and 100 s. During this time, the temperature reduced from around 3×10^{12} K to 10^9 K and the energy fell from about 1 GeV to 300 keV.

At the beginning of this epoch, the mean energy per particle was similar to the mass energy of a proton or neutron – about 1 GeV. This is why confinement of quarks became important from this time onwards. Before 10^{-5} s after the Big Bang, there had been sufficient energy available for up and down quarks to escape to distances significantly larger than the dimensions of a proton or neutron. After this time, no such escape was possible. Because a proton is composed of two up quarks and a down quark and a neutron is composed of two down quarks and an up quark, equal numbers of up and down quarks therefore led to an equal number of protons and neutrons emerging from this process. To recap on the contents of the Universe at this time, there were about ten billion photons, electrons, positrons, neutrinos, and antineutrinos for every single proton or neutron in the Universe.

Remember – the electrons and positrons had not yet mutually annihilated each other because the mass energy of an electron or positron is about 510 keV, and the mean energy per particle was still much higher than the ~1 MeV required to create a pair of them. So electrons and positrons were still in equilibrium with photons, and undergoing both annihilation and pair creation reactions at the same rate.

As soon as baryons had formed, weak interactions took over, with protons and neutrons rapidly changing into one another. Neutrons converted into protons by reacting with either positrons or electron neutrinos, while protons converted into neutrons by reacting with either electron antineutrinos or electrons. At the quark and lepton level, each of these reactions is a weak interaction because it involves quarks and leptons changing flavour. In each case, a W boson is involved.

With plenty of energy available, the transitions from neutron to proton and from proton to neutron proceeded at the same rate. Because there were as many neutrinos as electrons, and as many antineutrinos as positrons, the numbers of neutrons and protons in the Universe remained equal, at least initially. However, this situation did not continue. As noted earlier, the mass of a neutron is slightly higher than that of a proton. As a consequence of this, the reactions in which a proton converted into a neutron became slightly less likely to happen as the energy fell, because they required more energy than those in which a neutron was converted into a proton. As the Universe cooled, this difference in the rates of the two processes became more pronounced, and protons began to outnumber neutrons for the first time.

As the Universe cooled still further, another reaction became important for the neutrons and the protons: as you saw earlier, isolated neutrons decay into protons. This additional process, again governed by the weak interaction, added to the dominance of protons over neutrons in the Universe.

Once the Universe was 0.1 s old, the weak interactions in which neutrons and protons convert into one another became too slow, and neutrinos virtually ceased to have any further interaction with the rest of the Universe – ever! The ratio of protons to neutrons continued to rise as a result of neutron decay and was only halted when the neutrons became bound up in atomic nuclei where they became essentially immune from decay. As you saw earlier, if the typical lifetime of the neutron (about ten minutes) were much shorter than it in fact is, then all neutrons would have decayed into protons long before they could become confined inside nuclei.

It is worth pausing for a moment to try and comprehend just how short a time in the history of the Universe has elapsed at this point. Snap your fingers and that is about the same duration as the entire history of the Universe up to the instant now being discussed (about one-tenth of a second). Inflation, the breakdown of superunification, grand unification and electroweak unification, the formation of quarks and leptons, the subsequent decay of the massive particles, and the binding of up and down quarks into protons and neutrons – all occurred in the first tenth of a second after the Big Bang – literally in the blink of an eye...

When the Universe was about 10 s old, and the mean energy per particle was about 1 MeV, a final important event for the matter contents of the Universe occurred. The remaining primordial electrons and positrons mutually annihilated, producing yet more photons, but leaving the excess one-in-ten-billion electrons to balance the charges of the primordial one-in-ten-billion protons and ensure that the Universe has a net electric charge of zero.

16.6 PRIMORDIAL NUCLEOSYNTHESIS

As the temperature continued to decrease, protons and neutrons were able to combine to make light nuclei. This marked the beginning of the period referred to as the era of primordial nucleosynthesis (which literally means "making nuclei"). This next important phase of the Universe's history lasted just a few minutes from an age of about 100 s to 1000 s. The temperature at this time fell from 10^9 K to 3×10^8 K, implying an average energy per particle in the range 300 keV to 100 keV.

You have already encountered a set of nucleosynthesis reactions in Chapter 6, namely the proton–proton (pp) chain that occurs in the core of the Sun and other low-mass stars. However, the first step in the pp chain, in which two protons react together to form a deuterium nucleus, is a very slow process and did *not* occur to any great extent in the early Universe. On average an individual proton in the Sun must wait more than ten billion years before such a reaction happens, and there was simply not enough time in the early Universe for this to contribute significantly to nucleosynthesis.

The broad details of primordial nucleosynthesis were first explained in a 1948 publication by Ralph Alpher, Hans Bethe, and George Gamow. This paper is perhaps the most famous "joke" in all astrophysics. The work was actually done by PhD student Ralph Alpher and his supervisor George Gamow, but Gamow thought it would be amusing to add the name of the eminent nuclear scientist Hans Bethe to the author list, so that it could be read as "alpha, beta, gamma."

Instead of the pp chain reaction, the first such reaction to become energetically favoured in the early Universe was that of a single proton and neutron combining to produce a deuterium ("heavy hydrogen") nucleus, with the excess energy carried away by a gamma-ray photon. This reaction to produce deuterium does not occur as part of the pp chain in the Sun, as the Sun contain no free neutrons.

At high temperatures (greater than 10^9 K), there were a lot of high-energy photons in the Universe so deuterium nuclei were rapidly broken down. However, as the temperature fell below 10^9 K when the Universe was about 100 s old, deuterium production was favoured. Virtually all the remaining free neutrons in the Universe were rapidly bound up in deuterium nuclei, and from then on other light nuclei formed.

First, two deuterium nuclei may react together to form a nucleus of tritium with the ejection of a proton. Tritium is an even heavier isotope of hydrogen containing one proton and two neutrons in its nucleus. The tritium nucleus immediately reacts with another deuterium nucleus to form a nucleus of helium-4 with the emission of a neutron. The proton and neutron produced in the two reactions above can recombine to form another deuterium nucleus, so the *net* result of this set of reactions is that two deuterium nuclei are converted into a single nucleus of helium-4.

Other more massive nuclei were also made as deuterium nuclei reacted with protons to make nuclei of helium-3. These can then either react with other helium-3 nuclei to make helium-4 plus more protons or with nuclei of helium-4 to make beryllium-7. Nuclei of beryllium-7 are unstable and immediately capture an electron to form lithium-7 with the emission of an electron neutrino. Lithium-7 nuclei can react further with a proton to create nuclei of beryllium-8, but these too are unstable and immediately split apart into a pair of helium-4 nuclei. The end products of the four reactions are nuclei of helium-3, helium-4, and lithium-7, with the vast majority ending up as helium-4.

Nuclei with a mass number greater than seven did not survive in the early Universe. This is because there are no stable nuclei with a mass number of eight. The binding energy of a beryllium-8 nucleus is −56.5 MeV, which is slightly *more* than the binding energy of two helium-4 nuclei (−28.3 MeV each). Consequently, it is energetically favourable for beryllium-8 nuclei to split apart into two helium-4 nuclei, releasing 0.1 MeV of energy. The reactions that by-pass this bottleneck take much longer than the few minutes that were available for nucleosynthesis at this time. (Remember, the timespan here was around 15 minutes when the Universe had an age of between 100 and 1000 s.) Before more advanced reactions could occur, the Universe cooled too much to provide the energy necessary to initiate them.

The ratio of protons to neutrons had, by this time, reached about seven protons for every one neutron. Because the neutrons were bound up in nuclei, they no longer decayed, and the ratio remained essentially fixed from here on for the remainder of the history of the Universe. The vast majority of the neutrons ended up in nuclei of helium-4. Only very tiny fractions were left in deuterium, helium-3, and lithium-7 nuclei, because the reactions to produce them were far more likely to continue and produce helium-4 than they were to halt at these intermediate products. In fact, the actual proportion by mass of helium-4 that is predicted to have come out of the Big Bang according to the most recent calculations is 24.8%. The other 75.2% of the mass of nuclei created in the early Universe was virtually all hydrogen. The proportions by mass of deuterium and helium-3 were only about 1 part in 40,000 and 1 part in 100,000 respectively, and the proportion by mass of lithium-7 was about 1 part in 2 billion.

By the time the Universe had cooled to a temperature of about 3×10^8 K after 1000 s, the particles had insufficient energy to undergo any more reactions. The era of primordial nucleosynthesis was at an end, and the proportion of the various light elements was fixed. The rates of reaction to form helium and the other light elements have been calculated and the abundances predicted may be compared with the abundances of these nuclei that are observed in the Universe today. There is close agreement between theory and observation. In fact, this correspondence between the theoretically predicted abundances of the light elements and the observed abundances in the Universe today is the third major piece of evidence, alongside the cosmic microwave background and the Hubble expansion, in favour of the hot Big Bang model for the origin of the Universe.

So, at an age of around 1000 s, the Universe reached a state where its matter constituents were essentially as they are today. There are about ten billion photons for every baryon (proton and neutron), and about seven protons and electrons for each neutron. Neutrinos and antineutrinos continue to travel through the Universe unhindered by virtually anything they encounter.

16.7 STRUCTURE IN THE UNIVERSE

As the Universe cooled still further, nothing much happened for a few hundred years (between 1000 s and 10^{10} s). As the mean energy per particle fell below a few tens of electronvolts, electrons began to combine with nuclei to form neutral atoms for the first time. The final phase of the Universe's history that we consider is therefore from an age of a few hundred years to a few billion years. The temperature during this time range fell from 10^5 K to 3 K and the energy per particle dropped from about 30 eV to 10^{-3} eV.

Gradually, as electrically neutral matter accumulated, gravity began to take over as the dominant force operating in the Universe. Slight variations in the amount of matter and radiation in different regions meant that matter began to gather together into slightly denser clumps. These clumps provided the seeds from which clusters of galaxies later grew.

By the time the Universe had cooled to a temperature of 3000 K, about 300,000 years after the Big Bang, the mean energy of the photons had fallen to about 1 eV, and most of the matter in the Universe was in the form of neutral atoms. This was the trigger for another significant change in the behaviour of the Universe. The background radiation photons – those 10 billion photons for every particle left over from the annihilation epoch (matter and antimatter reactions) – interacted for the last time with matter in the Universe. When hydrogen atoms are in their ground state, photons with an energy of at least 10 eV are required in order to excite them to even the next energy level. So from this point on in the history of the Universe, the general sea of photons was no longer absorbed by matter. After this time, the background radiation simply expanded freely with the Universe, cooling as it did so, until it is observed today redshifted by a factor of 1000 into the microwave part of the spectrum as the cosmic microwave background (CMB).

When the CMB radiation is observed today, very slight irregularities are observed in its temperature and intensity. In part, these reflect slight differences in the matter distribution of the Universe at the time of the last interaction between the background photons and atoms. At the time of the discovery of these irregularities by the COBE satellite (Chapter 10), they were described as "wrinkles in the fabric of spacetime." On larger angular scales, above about 10 degrees across, the fluctuations in the CMB across the sky are due to the Sachs–Wolfe effect. This is caused by CMB photons undergoing gravitational redshift as they emerge from the strong gravity

FIGURE 16.3   The Hubble Ultra Deep Field. (Credit: NASA/STScI.)

in the vicinity of large clumps of matter. As a result, some regions of the sky will appear slightly cooler, by tiny fractions of a degree, as the spectrum of the CMB is shifted to longer wavelengths.

At some time after the last interaction of matter and radiation, but before the Universe was a billion years old, the first galaxies formed. As noted above, galaxies are thought to have formed as matter accumulated around slightly denser clumps, which themselves originated in random fluctuations that were present on a tiny scale in the first moments of the Universe's existence.

The most likely scenario for the formation of many galaxies is a so-called "bottom-up" merger process. The first large structures to form in the Universe had masses of around a million times that of the Sun. Over the next few billion years, many of these protogalaxies then merged, as a result of gravitational interactions between them, to form many of the galaxies that are seen in the Universe today. These have typical masses of a few hundred billion times that of the Sun. When astronomers study distant regions of the Universe, they are of course effectively looking back in time, because the light from the objects being observed has taken billions of years to reach us. So an image such as the Hubble Ultra Deep Field provides a window on the early Universe where protogalaxies may be seen in the process of merging to form larger structures (Figure 16.3). Many of the galaxies seen in that image are billions of light-years away, so are being seen as they were billions of years ago.

17 The Universe Today

This chapter finally reaches the point where a survey may be made of the Universe as it appears today, such that you can understand the processes that are observed to take place in it. However, as noted at the end of the last chapter, it is important to be careful about what is meant by the Universe "today." The problem is that, when observing the far distant Universe, astronomers are also looking far back in time, because the light and other radiation from those distant objects may have taken most of the age of the Universe to reach the Earth.

The chapter therefore begins by considering the largest scale structures that are seen, the distant galaxies that are observed soon after they formed in the early Universe, and the energetic events and processes occurring within them. After this, the chapter moves on to consider the Milky Way and other galaxies in the local neighbourhood of the Universe. Moving ever closer to home it then considers the formation of stars and planets, and how stars change and evolve through their lives. Finally, a description is provided of the Sun and planets of the Solar System, before the chapter rounds off with an examination of planets around other stars and a consideration of that most universal of questions – is there life elsewhere in the Universe?

17.1 LARGE-SCALE STRUCTURE

On the largest scales, the Universe is composed of vast sheets, walls, and filaments of matter comprising superclusters of galaxies, with equally immense voids in between them. This largest scale structure is known as the cosmic web (Figure 17.1).

Our local filament of the cosmic web is known as the Pisces-Cetus Supercluster Complex and is around 300 Mpc long by 50 Mpc wide, with a mass of a billion billion times that of the Sun. It includes the Virgo Supercluster of galaxies which in turn contains the Local Group of galaxies, including the Milky Way. Other named structures in the cosmic web include the Sloan Great Wall and the Hercules-Corona Borealis Great Wall, each of which is even larger than our local filament.

The Pisces-Cetus Supercluster Complex comprises about 60 clusters of galaxies, which may be grouped into the following five regions: the Pisces-Cetus Supercluster; the Perseus-Pegasus chain; the Pegasus-Pisces chain; the Sculptor region (including the Sculptor Supercluster and Hercules Supercluster); and the Laniakea Supercluster, which itself contains the Virgo Supercluster and also the Hydra-Centaurus Supercluster.

One way of studying distant clusters of galaxies makes use of a subtle effect that they have on the cosmic microwave background radiation, known as the Sunyaev–Zeldovich effect, named after the two Russian astronomers who described it in 1969. As low-energy photons from the cosmic microwave background pass through the

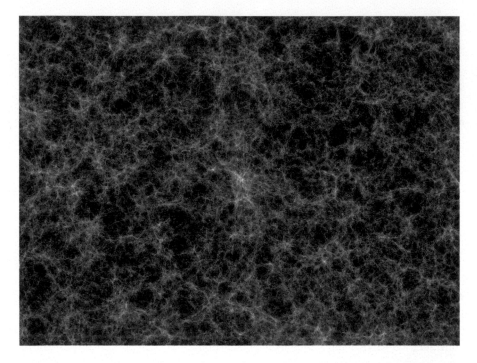

FIGURE 17.1   The cosmic web. (Credit: NASA/Volker Springel (Max Planck Institute for Astrophysics) et al.)

hot gas surrounding clusters of galaxies, they can undergo inverse Compton scattering from the electrons there and so gain a small amount of energy. This makes the temperature of the CMB in the directions of clusters of galaxies appear slightly hotter than it would otherwise be. Importantly, the size of the temperature shift does not depend on the redshift of the cluster of galaxies, which means that distant clusters at high redshift can be detected just as easily as nearby ones at low redshift (Figure 17.2).

Using such observations, it turns out that the Virgo Supercluster is merely one of around ten million superclusters of galaxies in the observable Universe. Within the Virgo Supercluster are at least 100 galaxy groups and clusters within its 33 Mpc diameter, with a total mass of about a million billion times that of the Sun. Its component groups include the Virgo cluster of galaxies, the Dorado, Leo, and Ursa Major groups of galaxies, the M81 and M101 groups of galaxies, and the Local Group of galaxies.

Coming finally closer to home, the Local Group of galaxies has two dominant members: the Milky Way and the Andromeda galaxy, plus dozens of smaller ones, within its 3 Mpc diameter. The total mass of the Local Group is around 2000 billion times that of the Sun, mostly contained within its two largest members (Figure 17.3).

Having considered the largest scale structures as whole, we now turn to consider individual galaxies and what they can tell us.

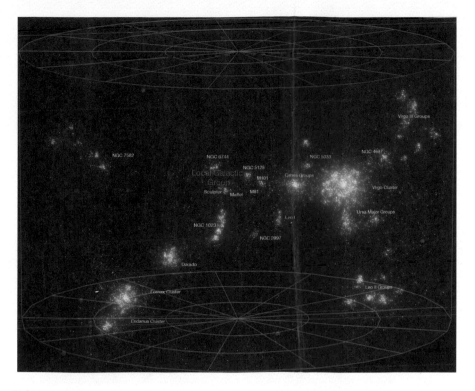

FIGURE 17.2   The Virgo Supercluster. (Credit: ESO/Andrew Z. Colvin/N. Bartmann.)

17.2   ACTIVE GALAXIES

When astronomers observe distant galaxies, many of them appear to have point-like, bright sources of emission in their centres. Indeed, when these objects were first discovered in the 1960s, they were named "quasi-stellar objects" (or QSOs) because they looked rather like stars. Many of them were also found to be strong sources of radio emission, and objects in this sub-category were referred to as "quasi-stellar radio sources" or quasars for short. Nowadays, QSOs are often referred to as quasars too, irrespective of whether or not they have strong radio emission. As mentioned earlier, it turns out that these are some of the most distant objects that astronomers can observe, with redshifts of up to 7 and look-back times of more than 10 billion years. The first quasar to be discovered goes by the catalogue name of 3C273 (number 273 in the third Cambridge catalogue of radio sources). It is also one of the nearest such objects and lies at a distance of about 750 Mpc away in the direction of the constellation of Virgo; it has a redshift of "only" 0.16.

As noted earlier, there is very good evidence that quasars are the energetic cores of distant galaxies, in which a supermassive black hole is accreting any matter that strays too close. Accretion is the process whereby objects accumulate matter from their surroundings or nearby objects, as a result of their intense gravitational fields.

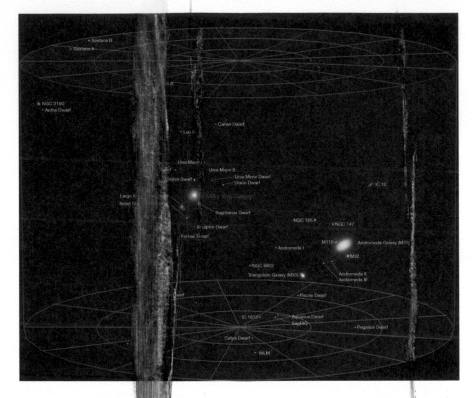

FIGURE 17.3   The Local Group of galaxies. (Credit: ESO/Andrew Z. Colvin/N. Bartmann.)

In doing so, gravitational potential energy is liberated, usually in the form of high-energy electromagnetic radiation.

Over the years, other types of galaxies displaying highly energetic processes in their central regions have also been discovered, and nowadays these various types are collectively referred to as active galaxies, with their energetic cores known as active galactic nuclei or AGN for short. Although different types of AGN are recognized, they are all believed to have certain features in common. The best model for an AGN is as follows.

In the very centre of an AGN sits a supermassive black hole, with a mass that is at least several hundred million (or even several billion) times that of the Sun. Surrounding the black hole is an accretion disc. This is a swirling disc of material that is orbiting around the black hole; as the material spirals inwards, it eventually gets swallowed by the black hole, passing beyond the so-called event horizon, from which there is no return. As material swirls within the accretion disc, frictional forces heat up the material to temperatures of millions of degrees, so that the disc glows brightly with optical, ultraviolet, and X-ray radiation (Figure 17.4).

Perhaps surprisingly, there is a limit to how luminous active galaxies, or any accreting system, can be. The luminosity emitted from the vicinity of the central source depends on the mass accretion rate, namely how much material falls

FIGURE 17.4   An artist's impression of the central regions of an active galaxy.

onto the black hole per unit time. A greater mass accretion rate leads to a higher luminosity – up to a point. The electromagnetic radiation that is liberated produces an outwardly directed radiation pressure. At some point, the pressure due to infalling material is exactly balanced by the outgoing radiation pressure, and so the mass accretion rate cannot increase any further. This situation is called the Eddington limit, named after the British astronomer Arthur Eddington who derived it. The luminosity of an object accreting at the Eddington limit is around 10^{31} W for every solar mass of the central object. As a result, a black hole with a mass of one billion times that of the Sun has a maximum luminosity of around 10^{40} W. The Eddington limit is an idealized situation for a spherical accretion flow composed entirely of hydrogen, but it provides a reasonable approximation in many cases.

Moving outward from the accretion disc, beyond this lies a region of gas and dust shaped something like a ring-donut, known as a torus. Between the torus and the accretion disc it is believed there exists a number of rapidly moving clouds of gas. These are each moving with a speed of up to several thousand kilometres per second in a random direction around the central regions of the AGN. If it were possible to measure the spectrum of an individual cloud, its emission lines would be seen to be red- or blue-shifted, with respect to the average speed of recession of the AGN as a whole, due to the Doppler effect arising from the cloud's motion. In fact, what is seen is an overall spectrum composed of the emission from *all* of the clouds at once, with a range of motion in all directions, so the result is that the emission lines in the optical spectrum are broadened significantly by the Doppler effect. Not surprisingly therefore, this part of the AGN is known as the broad-line region. Clouds moving with an average speed of order 3000 km s^{-1} will broaden the observed wavelength of an optical emission line by around 5 nm.

Further out still, beyond the torus, there seem to be a number of more slowly moving clouds of gas. Because these clouds are moving relatively slowly, with speeds of "only" several hundred kilometres per second, the spectral lines detected from them in the optical spectrum of an AGN are *not* broadened significantly by the Doppler effect. Not surprisingly again, this feature is known as the narrow-line region.

A particular class of active galactic nuclei are seen to exist in the heart of some spiral galaxies. These objects are known as Seyfert galaxies, named after the astronomer Carl Seyfert who first classified them in the 1940s. Interestingly there appear to be essentially two types of Seyfert galaxy: those in which both broad and narrow emission lines are seen in their optical spectra (Seyfert type 1) and those in which only the narrow emission lines are seen (Seyfert type 2). The idea is that the difference is simply down to the angle at which the AGN is viewed: Seyfert type 1s are seen close to face-on so both the broad-line region and the narrow-line region are seen clearly; but Seyfert type 2s are seen close to edge-on, so the inner broad-line region is hidden by the gas and dust torus, and only the outer narrow-line region is seen.

QSOs are now believed to be simply very luminous (and very distant) Seyfert galaxies. Indeed with the current generation of telescopes, faint host galaxies surrounding the point-like QSOs are now being detected. For many years, the only QSOs known had both broad and narrow emission lines in their spectra, so were referred to as type 1 QSOs, by analogy with Seyfert type 1s. However, in the last few years, some type 2 QSOs have been discovered displaying only narrow emission lines with obscured central regions. The discovery of such objects was a triumph for the theory that had predicted their existence.

Another division between types of AGN concerns whether or not they have strong radio emission. Sources that *do not* exhibit strong radio emission (known as "radio quiet" objects) include QSOs and Seyfert galaxies. Sources that *do* have strong radio emission (known as "radio-loud" objects) include quasars and the generically labelled radio galaxies. Radio galaxies display intense jets of material streaming away from their centres, each of which may end in an immense cloud known as a radio lobe. These radio jets and lobes may span distances up to megaparsecs in size, but the jets themselves are believed to originate close to the central black hole, emerging perpendicular to the accretion disc. Material in the jets travels at a large fraction (say, 50%) of the speed of light and so carries a vast amount of energy. If a radio-loud AGN is seen such that the view is essentially looking close to the axis of one of the jets, it is thought that what is observed is a quasar, and in this case the jets and radio lobes are not visible. The host galaxies of these radio-loud objects seem to be giant elliptical galaxies, rather than spirals.

Because all types of AGN can be seen at vast distances away and are therefore seen in the early history of the Universe, it is possible to use them to probe the evolution of the Universe with time. What is clear is that *more* AGN are seen in the past than there are today. In fact, the number density of AGN seems to have reached a maximum at a time in the past corresponding to a redshift of between 2 and 3, which equates to about 10 billion years ago. The number of them has been sharply declining ever since. There are two possibilities to explain this. The first possibility is that a small fraction of all galaxies contain AGN but that the luminosities of these AGN evolve with time so that at certain times in the past more of them are seen (this is called luminosity evolution). The second possibility is that virtually all galaxies contain AGN but, at any given time, most of them are inactive, or dormant, and the fraction of them that are active changes with time (this is called density evolution). In fact, various observations seem to suggest that a combination of the two effects has occurred, and luminosity-dependent density evolution describes the likely picture.

So where have all the AGN gone? It's clear that most large galaxies observed in the nearby Universe have supermassive black holes in their centres. It is possible, and perhaps likely, that many of these are the extinct remains of AGN, and that the active galaxy phenomenon may be seen as an energetic adolescent phase in the life cycle of many galaxies.

17.3   THE MILKY WAY AND OTHER GALAXIES

In the nearby Universe today, astronomers see broadly two types of large galaxy: elliptical galaxies with a shape that may be approximated by something like a squashed rugby ball, and spiral galaxies which are flattened discs containing spiral arms. There are also a large number of smaller, irregular galaxies, often seen as satellites to larger, more structured galaxies. The bottom-up merger scenario described at the end of the last chapter seems able to account for the formation of elliptical galaxies reasonably well, but the formation of spiral galaxies by this means is more of a problem: computer simulations show that the result of a merger between two similar-sized galaxies is always likely to be an elliptical galaxy, even if the original two galaxies were both spirals. Most spiral galaxies are probably therefore not the result of a bottom-up merger. Instead they probably formed by the in-fall of gas from the surrounding environment onto what were previously (smaller) elliptical galaxies. The in-falling gas can form a disc around the central bulge of the galaxy, which then develops into the spiral-arm pattern observed.

As previously noted, the galaxy in which the Sun and Solar System are located is known as the Milky Way and is a reasonably large spiral galaxy, often referred to as simply "the Galaxy." It has a mass of a few hundred billion times that of the Sun and is around 30 kpc in diameter, but its disc is only about 300 pc thick. In terms of structure it has a flattened disc-like shape, with a bulge in the centre. The disc of the Galaxy contains a number of spiral arms, which are regions where the density of gas, dust, and stars is greater. It is in the spiral arms where the on-going process of star formation is occurring, as discussed in the next section. In particular, many so-called open clusters are observed in spiral arms; these are groups of recently formed stars, each comprising typically a few thousand members. Distributed in a spherical halo around the Galaxy are several hundred objects known as globular clusters of stars (Figure 17.5). These are seen to be composed of around a million stars each and are all old stars, typically as old as the Galaxy itself, and so represent structures that formed along with the Galaxy, around 10 billion years ago.

In recent years it has become clear that the centre of the Galaxy contains a (mostly dormant) black hole, with a mass of 3.7 million times that of the Sun. Detailed

(a)   (b)

FIGURE 17.5   (a) A globular cluster and (b) an open cluster. (Credit: (a) NASA & ESA; (b) ESA/Hubble & NASA).

infrared images of the very central regions of the Galaxy over the past two decades have revealed a dozen or more individual stars orbiting around this invisible central object, allowing its mass to be calculated from their motion.

Amongst the nearest galaxies to the Milky Way are two dwarf irregular galaxies mentioned earlier, known as the Large Magellanic Cloud (LMC) and the Small Magellanic Cloud (SMC). They are only visible from the southern hemisphere of the Earth and are supposed to have been first brought to the attention of European science by voyagers who accompanied Ferdinand Magellan on his circumnavigation of the globe in the 16th century, although they were certainly known to Arabic astronomers centuries earlier. The LMC lies at about 50 kpc away and has a mass of around 10 billion times that of the Sun, whilst the SMC is around 60 kpc away and has a mass of about 7 billion times that of the Sun.

The largest neighbouring galaxy to the Milky Way is the most distant object that can be seen in the night sky with the naked eye. As noted earlier, it is known as the Andromeda galaxy (after its location in the constellation of the same name) or M31 (as the 31st object in the catalogue of "nebulae" compiled by Charles Messier in the late 18th century) (Figure 17.6). The Andromeda galaxy is a large spiral galaxy, almost 800 kpc away. It has a mass that is probably somewhat larger than the Milky Way, and it is a little larger in extent too.

The Andromeda galaxy and the Milky Way are in fact approaching one another with a speed of about 100 km s^{-1} and will collide in around 4.5 billion years' time. When this collision happens however, the two galaxies will largely simply pass through each other without any direct collisions between their stellar components. The gravitational forces will probably disrupt the spiral structures of each galaxy though and may result in the merged object resembling a giant elliptical galaxy. It is also likely that the resulting compression of gas clouds will cause bursts of triggered star formation, and the subject of the formation of stars is what we turn to next.

FIGURE 17.6   The Andromeda galaxy.

17.4   THE FORMATION OF STARS AND PLANETS

Having briefly summarized the large-scale structure that is seen in the Universe, it is now time to return to consider the processes that are seen to be on-going in the Milky Way. Within the first galaxies that formed, stars soon condensed out of the gas to become dense enough for nuclear reactions to start within their cores. But star formation was not a one-off process; star formation is still occurring today in galaxies throughout the Universe, including right here in the Milky Way.

The nearest star-forming region to the Earth is that which is observed as the Orion nebula and is about 400 pc away (Figure 17.7). There, a vast cloud of dust and gas, about 8 pc across, is in the process of collapsing to form newly born stars. The process probably began just a couple of million years ago, and the youngest stars seen there today may be only a few tens of thousands of years old.

Simply speaking, a cloud of gas and dust will spontaneously begin collapsing if its total energy is negative – that is to say if its kinetic energy is less than its gravitational potential energy. The kinetic energy of a cloud depends on its temperature and the number of particles it contains, whilst the gravitational potential energy of a cloud depends on its mass and size. Putting these facts together enables a lower mass limit to be derived, above which a cloud with a given temperature and size will collapse. This is known as the Jeans mass after James Jeans who first derived it in 1902. Cooler clouds will generally be more prone to collapse than hotter clouds because they have less kinetic energy, and dense clouds will be more prone to collapse than

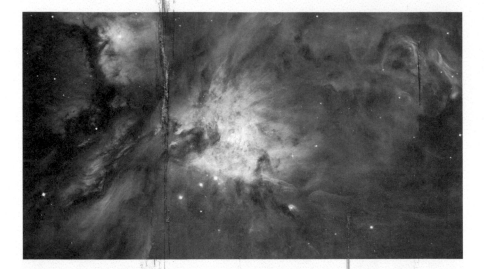

FIGURE 17.7   Star formation in the Orion nebula. (Credit: NASA/ESA/STScI/AURA/The Hubble Heritage Team.)

diffuse ones because they have greater gravitational potential energy. Clouds may also be triggered to collapse through some external influence, such as a nearby supernova explosion, or a collision between two clouds.

In the interstellar medium (the space between the stars), astronomers do indeed see examples of cool, dense molecular clouds that are prone to collapse. The clouds are mostly composed of hydrogen molecules, with some helium, and trace amounts of other elements. You should be aware, however, what astronomers mean when they use words like "cool" and "dense" … these molecular clouds are typically at temperatures of only 10 degrees above absolute zero and consist of perhaps a billion molecules per cubic metre. That may sound very dense, but it corresponds to a mass of only around one-millionth of a billionth of a gram per cubic metre which is far more tenuous than the best vacuum that can be achieved in a laboratory on Earth. The difference is, of course, that "dense" molecular clouds in space are vast in extent (typically several parsecs across) and therefore may contain a mass of thousands of times that of the Sun.

As clouds collapse in the way described above, they will break up into smaller and smaller regions as they become denser, which in turn become prone to collapse on their own as they in turn satisfy the Jeans criterion. For a given cloud, this process continues for around a hundred thousand years, at which point fragments of cloud with a mass comparable to that of the Sun have typically formed. The cloud fragment will have heated up by this point, as gravitational energy has been converted into thermal energy of the molecules within it. Eventually what forms may be termed a protostar, and it becomes opaque to the electromagnetic radiation (chiefly infrared light) that is liberated within it. This allows it to heat up further, which increases the pressure within, and the collapse eventually halts. By the time an equilibrium balance has been reached between gas pressure acting outwards and gravitational forces acting inwards, the internal temperature of the object has reached typically 30,000 kelvin.

Protostars are generally shrouded by dust remaining from the collapse, but astronomers can observe them using infrared observations. At this stage in their lives, protostars are surrounded by a vast, flattened disc of material perpendicular to the axis of rotation of the object. Often, bipolar outflows, or jets, are observed emanating along the rotation axis too. Eventually, over the course of a hundred million years or so, the shrouding dust will disperse and the disc of gas may condense into a system of planets, revealing a fully fledged star in the centre.

As just noted, planets are believed to form in the proto-planetary disc surrounding young proto-stars. Grains of dust within this disc will tend to settle towards its mid-plane, where they can come into contact with other grains, and gradually coagulate into larger particles. In the inner parts of the disc, rocky particles (silicates) can condense out; but further away from the protostar, it will be cold enough for ices such as water, ammonia, and methane to condense too. The snow line is the nominal boundary between the two regions.

Within a few hundred thousand years, the coagulating particles will have grown into planetesimals up to a few kilometres in size. From here on, the gravitational pull of these planetesimals is large enough to attract surrounding material, allowing a few objects to undergo runaway growth to form planetary embryos with masses around a tenth that of the Earth. Inside the snow line, terrestrial planets probably form by a series of collisions between multiple planetary embryos, whilst gas giant planets are able to form in the outer regions of the protoplanetary disc (beyond the snow line) as planetary embryos grow large enough there to attract gas directly and form the extended envelopes that are seen today in planets such as Jupiter and Saturn in the Solar System.

Having seen how stars form, the next stage of the journey is to consider how they evolve.

17.5   THE EVOLUTION OF STARS

Stars are "born" when their cores become hot and dense enough to initiate the hydrogen fusion reactions that you read about in Chapter 6. Fusion reactions release energy and so prevent stars from collapsing further due to gravity. It is perhaps counter-intuitive to realize that stars undergo nuclear fusion in their cores *because* they are hot; they are *not* hot because they undergo nuclear fusion in their cores. They are hot because they have collapsed from much larger clouds of gas, and in fact nuclear fusion *prevents* the cores of stars from getting any hotter.

It turns out that the Sun is a fairly typical or average star. For this reason, masses, radii, and luminosities of other stars are often conveniently expressed in terms of the mass, radius, and luminosity of the Sun itself. One solar mass (written as $M_\odot$) is about 2×10^{30} kg, one solar radius ($R_\odot$) is about 7×10^5 km, and one solar luminosity ($L_\odot$) is about 4×10^{26} W.

17.5.1 LIFE ON THE MAIN SEQUENCE

The first stars in the Galaxy were probably born over 10 billion years ago. Deep within these stars, hydrogen was converted into helium via the proton–proton chain, releasing energy as electromagnetic radiation into the Universe. The same reaction powers the cores of low-mass stars, like the Sun, today.

In later generations of stars, particularly those of higher mass which already incorporated some heavier elements such as carbon and oxygen into their initial composition, hydrogen fusion can occur via another route, called the CNO cycle, first worked out by Carl von Weizsäcker and independently by Hans Bethe in 1939 (Figure 17.8). Starting with a nucleus of carbon-12, protons can be added (followed by a couple of beta-minus decay reactions) to produce successive nuclei of nitrogen-13, carbon-13, nitrogen-14, oxygen-15, and nitrogen-15. When this last nucleus absorbs a final proton, it spits out a helium-4 nucleus and so transforms back into the original carbon-12 nucleus, recovering the initial catalyst. The net result is the conversion of four protons into a helium-4 nucleus, just as with the proton–proton chain.

Whilst stars are converting hydrogen to helium, they can be said to be in the main phase of their lives, and the phase that lasts the longest. The two key observable quantities for any star are its luminosity (how much energy it emits) and its temperature (how hot it is). These two quantities may be measured for a large number of stars and the results plotted on a graph of luminosity versus temperature, known as a Hertzsprung–Russell (HR) diagram, after the two scientists who independently devised it (Figure 17.9). When this is done, an interesting result is seen: most stars lie

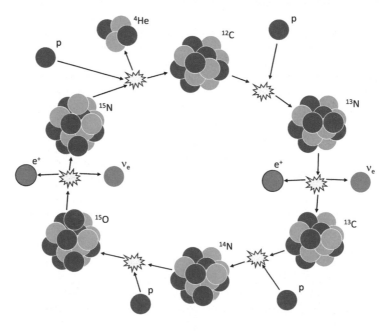

FIGURE 17.8   The CNO cycle.

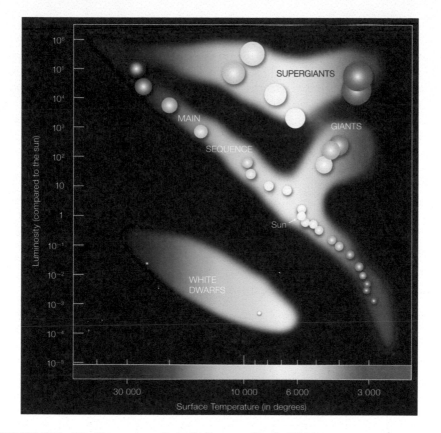

FIGURE 17.9   The Hertzsprung–Russell diagram. (Credit: ESO.)

along a single track, known as the main sequence, running from hot luminous stars at the top left, down to cool faint stars at the bottom right.

This track delineates where stars sit during their hydrogen-burning lifetimes, whether they make use of the proton–proton chain or the CNO cycle to do so. Not surprisingly, the hot luminous stars are also the most massive and largest, whilst the cool faint stars are the least massive and smallest. The most massive stars at the top of the main sequence have masses up to 100 $M_\odot$, radii of perhaps 15 $R_\odot$, and luminosities of up to a million $L_\odot$. At the other extreme, the least massive stars at the bottom of the main sequence have masses as small as 0.1 $M_\odot$, radii of less than 0.1 $R_\odot$, and luminosities as small as one ten-thousandth of $L_\odot$.

Stars on the main sequence are conveniently classified by astronomers on the basis of their spectra into one of seven types, referred to by the letters O, B, A, F, G, K, or M running from top left to bottom right. The sequence was originally alphabetic based on the strength of hydrogen lines in stellar spectra but was later reordered (and some classes merged) when temperature was recognized as their defining characteristic. O- and B-type stars are the hottest blue and blue-white stars, and main sequence examples include the stars theta[1] Orionis C (in the heart of the Orion nebula) and Eta Aurigae. The A-type stars (like Vega) are white, F-type stars (like Procyon) are yellow-white, and G-type stars (like the Sun or Alpha Centauri A) are yellow. Cooler than the Sun, the K-type stars (such as Epsilon Eridani) are orange, and the M-type stars on the main sequence are known as red dwarfs. Examples of the latter include Proxima Centauri and Barnard's star. Fainter, smaller, and cooler still are the so-called brown dwarf stars which are not massive enough to have initiated hydrogen fusion in their cores. They are sometimes referred to as L-type or T-type stars and may be plotted at the extreme lower right of the main sequence. Named examples of these classes include GD 165B and Gliese 229B.

The Sun therefore sits somewhere below the middle of the main sequence. Just like some movie stars, the biggest and brightest stars live fast and die young. The most massive ones run out of hydrogen in their cores most rapidly – perhaps over only a few tens of millions of years – whilst low-mass stars can take tens of billions of years to undergo any significant change. The main sequence lifetime of the Sun is about 10 billion years, and it's currently about half-way through that life.

17.5.2 Post-Main Sequence Evolution

As stars grow older, their cores contract and grow hotter as the hydrogen fuel in the core is gradually used up and converted to helium. For stars with less than around ten times the mass of the Sun, this causes their outer layers to expand and cool, and the star becomes a bloated red giant. During this process a star moves away from the main sequence on the HR diagram towards what is known as the (first ascent) red giant branch (RGB). Stars on the RGB appear cooler (redder) but more luminous (brighter) than they were on the main sequence. They can have radii up to 100 times that of the Sun. At this stage in a star's life, hydrogen fusion continues in a shell surrounding the inner core of processed helium. Examples of red giant stars include Pollux, Arcturus, and Aldebaran.

From here on, further nuclear fusion reactions can occur in stars at the higher end of the mass range. The details of these processes were published in 1957 by Margaret and Geoffrey Burbidge, William Fowler, and Fred Hoyle in a paper called "Synthesis of the elements in stars" and referred to generally as "B^{2}FH." In this work, they showed how carbon, oxygen, and many other elements were formed and described the so-called s-process and r-process which are described below.

As long as a star is more massive than about half the mass of the Sun, the core will eventually become hot enough for helium fusion to occur. The first such reaction to occur is known as the triple-alpha process in which three nuclei of helium-4 (also known as alpha particles) fuse together to form a nucleus of carbon-12 (with an atomic number of 6). Later, another helium-4 nucleus may also be added to produce oxygen-16 (with an atomic number of 8). The onset of stable helium fusion in the core moves a star to a new location on the HR diagram, somewhat fainter and hotter than the tip of the RGB. Whilst undergoing core helium burning, stars sit in a region of the HR diagram known as the horizontal branch. The branch is "horizontal" because all stars in this phase have roughly the same luminosity, but slightly different temperatures, depending on the amounts of heavy elements (other than hydrogen and helium) that are present. The pulsating RR Lyrae stars, mentioned earlier as standard candles, are core helium-burning stars on the horizontal branch.

In fact, pulsating stars are found across the HR diagram in a constrained region known as the instability strip. This runs from the middle of the main sequence towards the upper right of the HR diagram, and encompasses RR Lyrae stars, Cepheid variables, and other types too. The reason for its existence is that stars with certain combinations of luminosity and temperature are prone to an instability in their outer layers caused by the fact that the opacity of ionized helium varies with temperature. When a star is compressed by gravity, a layer within the star containing singly ionized helium heats up and becomes doubly ionized. This causes it to become opaque, trapping light inside. In turn, this further heats the gas inside the star and increases its pressure. The high-pressure gas eventually expands and cools, the doubly ionized helium re-combines with electrons to become singly ionized again, and in doing so becomes transparent. Light can now escape, so the gas cools and the pressure falls. When the pressure falls, the star is once again compressed by gravity and the process repeats cyclically. This is known as the kappa mechanism, because the Greek letter kappa is the symbol for the numerical value of the opacity of a gas. During this cycle, the star grows and shrinks in size, becoming brighter and fainter as it does so in a characteristic manner that may be observed.

Returning to the evolution of core helium burning stars on the horizontal branch, eventually the core will become depleted in helium too, and the star may begin another movement across the HR diagram. It becomes larger, cooler, and more luminous again, this time ascending the asymptotic giant branch (AGB), as it begins shell helium burning via the triple-alpha process around an inert core. This second red giant branch is referred to as "asymptotic" because it parallels the earlier first ascent giant branch.

During the AGB phase, some further nuclear reactions can also occur in these stars. In the helium-burning shell, nuclei of carbon-13 (left over from the CNO cycle) can fuse with helium to create oxygen-16 and release a free neutron. These neutrons can then add, slowly one at a time, to other nuclei to make even more massive elements. As more and more neutrons are added, some transform into protons, via beta-minus decay, and in this way massive (stable) nuclei can be created – all the way up the periodic table to lead and bismuth, with atomic numbers 82 and 83. This so-called s-process (where "s" stands for slow) is the means by which the most massive, non-radioactive nuclei that exist in the Universe are created. Elements such as silver, barium, tungsten, and mercury are predominantly formed in this way. They may be expelled into the wider Universe during this phase of a star's life by thermal pulses in which a star sheds some of its outer envelope in events that recur every tens to hundreds of thousands of years for an individual star.

For the most massive stars, with initial masses at least ten times that of the Sun, the evolutionary tracks are rather different. These stars are already very luminous, even when on the main sequence, and evolve to cooler temperatures as they age without changing much in luminosity. They become supergiant stars with radii up to 1000 $R_\odot$ and can track back and forth on the HR diagram as different phases of nucleosynthesis begin and end. Their evolution is very rapid however, and such stars

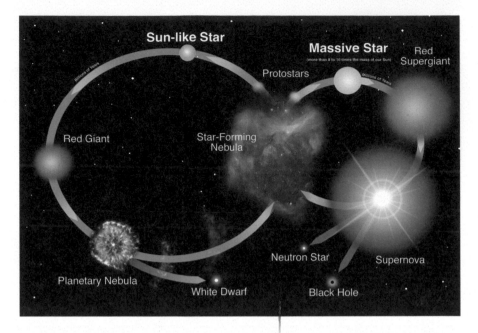

FIGURE 17.10 The evolution of stars. (Credit: NASA and the Night Sky Network.)

are also relatively rare, so their evolution is not so well charted as that for lower mass stars described above. They are nonetheless extremely bright and so readily noticed. Examples of supergiant stars include Rigel and Deneb (both blue supergiants), as well as Betelgeuse and Antares (both red supergiants).

Eventually though, nuclear fusion in stars comes to an end, and what happens subsequently is considered next (Figure 17.10).

17.5.3 STELLAR DEATH

As low-mass stars run out of nuclear fuel, they will eventually shed their outer layers to form a diffuse cloud of metal-enriched gas, which becomes visible as a so-called planetary nebula (Figure 17.11). The core of the star that is left behind will collapse to form a dense, compact object known as a white dwarf, supported against further collapse by an effect of quantum physics known as electron degeneracy pressure. Essentially, the Pauli exclusion principle (which you met in Chapter 4) prevents the electrons from being packed together any more closely, and so is able to support the weight of the outer layers of the white dwarf.

The Sun and other stars up to about eight times the Sun's mass will end up as white dwarfs composed of carbon and oxygen. A white dwarf is a very dense object. It has a mass similar to that of the Sun, but a radius comparable to that of the Earth. Several tonnes of white dwarf material would comfortably fit in a matchbox! The first white dwarf discovered was 40 Eridani B, the nearest one to us is Sirius B, and the first isolated such star found was van Maanen 2.

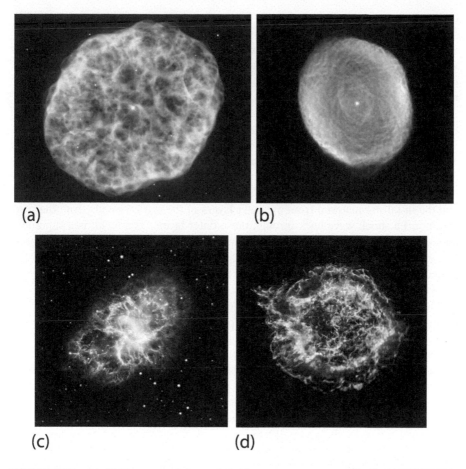

FIGURE 17.11   (a), (b) Planetary nebulae and (c), (d) supernova remnants. (Credit: (a) ESA/Hubble and NASA, acknowledgement: Marc Canale; (b) NASA and The Hubble Heritage Team (STScI/ AURA), acknowledgement: Raghvendra Sahai (JPL) and Arsen R. Hajian (USNO); (c) NASA, ESA, NRAO/AUI/NSF and G. Dubner (University of Buenos Aires); (d) NASA/CXC/SAO.)

The more massive the star, the hotter is its interior, and the more massive the elements that can be produced by nuclear fusion reactions. In stars that are more massive than about 8 solar masses, further helium fusion reactions can produce ever heavier nuclei, such as neon-20 (atomic number = 10) and magnesium-24 (atomic number = 12). Some of these stars will end their lives as oxygen–neon–magnesium white dwarfs, via a similar process to that described above, but such white dwarfs will always be less massive than about 1.4 times the mass of the Sun. This is known as the Chandrasekhar limit and marks the limit beyond which electron degeneracy pressure is unable to support a white dwarf against further collapse.

In stars that are more massive still, heavier elements such as silicon-28 (with atomic number = 14) and sulphur-32 (with atomic number = 16) can also form in the stellar core. But there is a limit to how far nuclear fusion can go. As you saw in Chapter 6, when four protons are converted into a nucleus of helium-4, the products have a lower mass than the reactants. This mass difference is liberated as energy. Similar mass reductions apply for reactions to produce all the elements up to those with atomic numbers in the range of about 26 to 28, namely iron, cobalt, and nickel which sit at the lowest point of the binding energy per nucleon graph. However, for nuclear fusion reactions beyond this, more energy must be put into the reactions than is released from them, so these are not viable.

But what happens to these massive stars? When the core is largely composed of iron, they have no further source of energy available. The outer layers fall inwards, squeezing the centre of the star down until it has a density comparable to that of an atomic nucleus. In the process, protons and electrons in the iron atoms of the star's core are forced to combine to make neutrons – effectively converting the entire core of the star into an incompressible ball of neutrons. The collapse halts – suddenly – and the in-falling material from the outer layers of the star rebounds from the core, setting off a shock wave back through those outer layers. The result is a supernova explosion, in which 90% or more of the star's mass is thrown violently out into space. The expelled cloud of material will glow for a few thousand years and be visible as a supernova remnant, before it fades away. It contains the nuclear processed material from the outer layers of the progenitor star.

This is because, in the final moments of its life before undergoing a supernova explosion, a massive star has one final surprise left. The immense temperatures and pressures created during the supernova cause electrons and protons in the star's outer envelope to react to form huge numbers of free neutrons. These neutrons enable elements to be built *beyond* the lead and bismuth limit. In this r-process (where "r" stands for rapid), precious metals such as platinum and gold are formed as well as naturally occurring radioactive elements, such as radium, thorium, and uranium. They are dispersed into the Universe by the explosion.

The star's core left behind will be revealed as a super-dense neutron star, with a mass of about one and a half to two and a half times that of the Sun contained within a volume that is only about 10 km across; a million tonnes of neutron star material would fit on a pinhead. Neutron stars are supported against further collapse by neutron degeneracy pressure – another consequence of the Pauli exclusion principle which limits how closely neutrons may be packed together.

Neutron stars also spin very rapidly, due to the process of conservation of angular momentum. In the same way that an ice skater will spin more rapidly if she pulls in her arms, so when a star collapses to a neutron star, it too will increase its rate of spin. The spin of a neutron star is apparent because many of them emit beams of radio waves from their magnetic poles. Just as the Earth's magnetic poles are offset from its geometric poles, the same thing occurs in a neutron star. So as a neutron star rotates, its radio beams will sweep across the sky, like the beam of a lighthouse. If a telescope happens to lie along a line of sight that intercepts that beam, a "pulse" of radiation will be seen every time the neutron star rotates.

The first such object was discovered in 1967 by British radio astronomer Jocelyn Bell and they were subsequently named pulsars. An alternative name initially suggested for these objects was "LGM", standing for "Little Green Men." However, an alien origin was soon discounted and pulsars were recognized as rotating neutron stars, which had been hypothesized to exist some 30 years earlier, soon after neutrons were discovered. Almost 3000 radio pulsars have now been identified, with rotation periods ranging from a few milliseconds to a few seconds. The most famous radio pulsar is the one sitting in the centre of the Crab Nebula supernova remnant (Messier object, M1). The Crab Pulsar spins 30 times a second and is the remains of an exploding star observed by Chinese astronomers in the year 1054.

For the very largest stars, the core is so massive that even a neutron star configuration is not stable. The Oppenheimer–Volkoff limit specifies that neutron degeneracy pressure can only support a neutron star with a mass less than about 2.5 times that of the Sun. With a core mass greater than this, still further collapse of the stellar core occurs and the resulting object is revealed as a black hole. A black hole is so called because it is so dense that not even light can escape from it. In theory the entire mass of a black hole is contained within an infinitely small volume of space, known as a singularity, and black holes therefore have an infinite density. In practice the singularity is always hidden from view within a boundary known as the event horizon. The size of this boundary, referred to as the Schwarzschild radius, is about 3 km per solar mass, so a 5-solar mass black hole has a Schwarzschild radius of only 15 km. The event horizon marks the dividing line beyond which nothing can escape from the black hole's pull.

White dwarfs, neutron stars, and stellar mass black holes are often collectively referred to as compact objects. Their immense gravity can influence other stars in their vicinity, as explained next.

17.5.4 Accreting Compact Binary Stars

Many stars are born as part of binary systems, and many of these binaries persist even after the death of one component and its transformation into a white dwarf, neutron star, or black hole. As part of the evolution of the star that is now a compact remnant, it will likely have undergone phases in which it expanded greatly (on the red giant and asymptotic giant branches of the HR diagram). During these phases it may have engulfed its companion star, if it was orbiting close enough, and as a result the two stars would have spiralled closer together due to frictional drag during the so-called common-envelope phase.

As a result, when the more evolved star has shed its outer layers and become a white dwarf, neutron star, or black hole, the emerging binary would have a much shorter orbital period of typically only a few hours, as the two stars are now much closer together than they were originally. In this configuration, the dense compact object can literally pull material from its companion in a process known as Roche lobe overflow accretion. The companion star is distorted into a "pear-shape" by the strong gravity of the compact object and material overflows from the tip (referred to as the inner Lagrangian point) into the deeper gravitational potential well of the compact object. Here it spreads into a flattened accretion disc (similar to that in active galaxies mentioned earlier) before the material empties onto the compact object itself.

These close, accreting compact binary stars are referred to as cataclysmic variable stars (CVs) if the accreting object is a white dwarf, or low-mass X-ray binary stars (LMXBs) if the accreting object is a neutron star or black hole (Figure 17.12). Examples of CVs include the stars U Geminorum, SS Cygni, and AM Herculis, whilst examples of LMXBs include Cygnus X-2, Aquila X-1, and Scorpius X-1. A variety of phenomena are observed in these systems, including outbursts that arise in the accretion disc surrounding the compact object, or on the surface of the compact object itself (apart from in black hole systems which have no surface on which matter can accumulate). Typically, their accretion discs are very hot – glowing in the ultraviolet or X-ray parts of the spectrum – and the outbursts too are often prominently seen in these energetic parts of the spectrum.

Even if the binary stars are quite well separated, accretion onto the compact object can still occur from the strong stellar wind of the companion star, in a process known as Bondi–Hoyle accretion. In high-mass X-ray binary stars (HMXBs), a neutron star or black hole accretes from the wind of a blue giant or supergiant star, while in symbiotic binary stars, a white dwarf accretes from the wind of a red giant. Such systems may have orbital periods of between several days and several years. Examples of HMXBs include Vela X-1, Cygnus X-1, and SMC X-1, whilst the prototype symbiotic star is Z Andromedae.

In even more extreme systems, both components may be compact stars. So-called ultracompact X-ray binary stars, with orbital periods as short as a few minutes, consist of pairs of white dwarfs or a neutron star and a white dwarf. In each case, the more massive component accretes from the less massive component, again giving

FIGURE 17.12 A compact star accreting material from its companion. (Credit: NASA/CXC/M. Weiss.)

rise to copious X-ray emission. The system with the shortest known orbital period is 4U 1820–30 in which a neutron star and a white dwarf orbit each other every 11 minutes.

Having explored these extreme end points to stellar evolution, we now return to the life cycle of a more sedate star – the Sun.

17.6   THE SOLAR SYSTEM

Following the death of a massive star, the star cycle repeats – but this time with a slight difference. Stars that formed after the first generation had lived and died had a richer source of raw material from which to build. A star like the Sun, or one of those in the Orion nebula, was formed in a galaxy that had already seen at least one generation of massive stars born, live, and die in supernovae explosions. The gas and dust from which the Sun formed, about 5 billion years ago, had therefore been enriched by heavier elements produced inside these earlier stars.

When it comes to categorizing chemical elements, astronomers have a very simplified view of the periodic table; they commonly refer to just hydrogen, helium, and metals, where the latter term encompasses *all* the other naturally occurring elements. This is because the Universe is predominantly composed of only the first two elements, and everything else makes up only a very small proportion. The metallicity of the Sun is about 2%, meaning that only 1/50 of its mass comprises elements other than hydrogen (comprising about 70% by mass) or helium (comprising about 28% by mass). Stars that are older than the Sun were typically born with even smaller metallicities, whilst younger stars will generally have a higher value of metallicity. Because the nebula from which the Solar System formed had this enriched composition, this led to the possibility of the formation of planets from the rubble left behind after the Sun was born.

The Earth itself formed from such debris. Every nucleus of carbon, oxygen, magnesium, and silicon found on the Earth and within living creatures was created inside the heart of an ancient star. Every nucleus of elements heavier than iron, such as strontium, zirconium, silver, and gold, was formed either from slow neutron capture in ageing stars, or by rapid neutron capture during the supernova explosions that mark their death.

17.6.1   THE SUN

The Sun itself is a very average star and as noted earlier, astronomers typically describe other stars' properties with reference to those of the Sun. The volume enclosed by the inner 25% of the Sun's radius is referred to as the core and is where nuclear fusion reactions occur (Figure 17.13). The core temperature of the Sun is almost 16 million kelvin; this is hot enough for electrostatic repulsion between protons to be overcome, and so initiate hydrogen fusion by the pp chain, as noted earlier. Overlaying the core is a region of ionized gas known as the radiative zone. X-ray photons released in the core are repeatedly scattered, absorbed, and re-emitted throughout the radiative zone over the course of millions of years, as they gradually

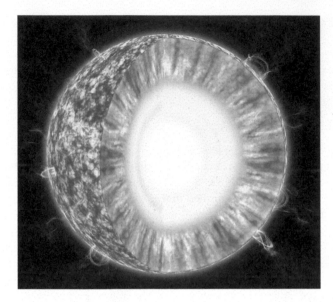

FIGURE 17.13   The structure of the Sun.

diffuse outwards. The outer region of the Sun, occupying the final 30% of its radius is the convective zone, where vast cells of gas rise and fall like water boiling in a pan, carrying energy to the surface.

The apparent surface of the Sun is known as the photosphere and marks the point beyond which the atmosphere of the Sun becomes transparent. This region has a temperature of about 5800 kelvin and is the source of the black-body spectrum that peaks in the visible part of the spectrum and gives the Sun its characteristic yellow colour.

Although the Sun may appear to be a tranquil and unchanging ball sitting in the sky, its surface is subject to violent eruptions and turmoil. Imaging the Sun in ultraviolet radiation or X-rays reveals vast loops of plasma arching around the Sun's complex magnetic field lines, lifting material way above the Sun's photosphere into space. Solar flares are brightenings of regions of the Sun's surface associated with ejections of clouds of material. The material thrown into space by these events can eventually reach the Earth a few days later, where the charged particles captured by the Earth's magnetic field result in displays of aurora for those near the polar regions.

17.6.2   TERRESTRIAL PLANETS AND ASTEROIDS

All the inner planets of the solar system are subject to solar weather as they intercept the "wind" of charged particles blown off the surface of the Sun. There are four terrestrial planets: Mercury, Venus, the Earth, and Mars, although the Moon and some of the larger satellites of the outer planets are also considered to be terrestrial-like bodies and have similar composition or structure in some cases (Figure 17.14).

FIGURE 17.14 The Sun and some constituents of the Solar System (not to scale).

Detail of the internal structure of Earth is provided by both direct and indirect evidence. Composition of the lithosphere, that is the crust and upper mantle, can be determined by examination of key rock types. The structure and composition of the deep mantle and core are revealed from the properties and response of seismic waves that pass through the planet. Similar studies of other terrestrial planets may be carried out from planetary landers or satellites.

Terrestrial planets retain internal primordial heat from processes that operated in the early stages of planetary evolution. This primordial heat includes that derived from the collision and assembly of planetary embryos, as well as that delivered to the surface by incoming impactors after the planet had assembled. Internal heat generation within terrestrial planets such as the Earth is mainly radiogenic heating as a result of radioactive decay of uranium-238, uranium-235, thorium-232, and potassium-40 in their silicate-rich mantle and crustal layers. The amount of radioactive decay was greater early in a planet's evolution because there would have been considerably more radioactive elements present.

The terrestrial planets have a variety of different atmospheres. Mercury and the Moon have no atmospheres to speak of, but Venus has an extremely thick atmosphere, dominated by carbon dioxide, producing a pressure almost a hundred times that of the Earth's atmosphere. Mars too has a carbon dioxide atmosphere, but here the pressure is only around half-a-percent that of the Earth's. Only on Earth, with its nitrogen/oxygen-rich atmosphere, are the temperatures and pressures thought to be conducive to the existence of "life as we know it."

None of the terrestrial planets' atmospheres are relics from their formation. Any primordial hydrogen and helium atmospheres (such as are seen in the giant planets) would rapidly have leaked away into space, as these planets are not massive enough to retain them. Instead, the atmospheres seen today are the result of out-gassing from the planets' interiors, releasing principally carbon dioxide and nitrogen. These more massive molecules can be retained by Venus, the Earth, and Mars, but cannot be retained by Mercury and the Moon, which are too small (and in Mercury's case, too hot as well). The differences between the atmospheres of the terrestrial planets then largely come down to water. On Venus, water is all vaporized, whereas on Mars, water is present as solid ice (in the polar caps or below the surface). On Earth, carbon dioxide is readily dissolved in liquid water, and the high abundance of oxygen arises due to the conversion of carbon dioxide into oxygen by plants during photosynthesis. The concept of a planetary habitable zone, where it is neither too hot nor too cold for liquid water to exist, is crucial when it comes to examining exoplanetary systems as potential habitats for life.

Beyond the inner terrestrial planets, and before reaching the outer giant planets, lies the majority of the asteroid belt. More than half a million asteroids have been discovered to date, but there are probably twice as many as this that are bigger than a kilometre in size, and numbers beyond counting that are smaller. The two tiny moons of Mars, Phobos and Deimos, are probably asteroids that strayed too close and got captured by the planet's gravitational pull. Also within the asteroid belt lies the dwarf planet Ceres, although the three next-largest asteroids (Pallas, Vesta, and Hygiea) might also be classified as dwarf planets too.

17.6.3 GIANT PLANETS AND THEIR SATELLITES

The giant planets of the Solar System, in order of their distance from the Sun, are Jupiter, Saturn, Uranus, and Neptune. Data on the interiors of the giant planets can be obtained from measurements of density, gravitational field, magnetic field, emitted heat, and atmospheric composition, often made from spacecraft fly-by or orbit.

Current models of Jupiter and Saturn distinguish five layers. The two innermost layers constitute a core of rocky and icy materials. This core is surrounded by layers that are mostly hydrogen and helium, and accounts for most of the planets' mass. The layer adjacent to the core in Jupiter and Saturn is predicted to contain hydrogen in a metallic state. The deep interiors of both Jupiter and Saturn are very hot (over 15,000 K in the case of Jupiter), but the temperatures of the outer layers fall to below 200 K.

Uranus and Neptune may not have a definite liquid or solid surface; they may have rocky cores but current models suggest that rocky and icy materials are not completely differentiated within their structures. Surrounding the core is a mantle of mainly icy materials and around this is a layer of mainly hydrogen and helium. Overall, these two planets are less dominated by hydrogen and helium than Jupiter and Saturn, and the layers are probably less differentiated in composition.

Although the atmospheres of the giant planets have hydrogen and helium as their major components, other molecules detected include methane and ammonia. Most of the molecules in the atmospheres are detected by infrared or ultraviolet spectroscopy but the Galileo probe also used mass spectrometry to obtain the relative abundances of molecules in the region of Jupiter's atmosphere it entered. The outermost cloud layer can be identified as ammonia on Jupiter and Saturn, but clouds of methane have been observed in the atmospheres of Neptune and Uranus.

Each of the giant planets is accompanied by a system of rings – although those of Saturn are the only ones substantial enough to be seen in a small telescope from Earth. The rings of Saturn are composed of countless particles of water ice, ranging in size from less than 1 cm to more than 1 m. The rings themselves extend up to 80,000 km from Saturn's cloud tops but are no more than 1 km thick.

The giant planets also have large numbers of satellites – over 170 between them at the time of writing. The four largest moons of Jupiter (Io, Europa, Ganymede, and Callisto), the largest moon of Saturn (Titan), and the largest moon of Neptune (Triton) range in size between roughly the size of the Moon and the size of the planet Mercury. If they were situated elsewhere in the Solar System they might each be classed as terrestrial planets. Saturn's other major satellites (Enceladus, Tethys, Mimas, Dione, Iapetus, and Rhea) and the major satellites of Uranus (Miranda, Ariel, Umbriel, Titania, and Oberon) are each between one-tenth and half the diameter of the Moon, comparable in size to the largest asteroids.

Many of these moons are interesting planetary bodies in their own rights. The tidal pull of Jupiter results in active volcanoes on Io, whilst Europa is covered by a thick ice-layer that may hide a liquid water ocean beneath, and Enceladus exhibits cryovolcanism, in which jets of ice crystals are spouted into space. Titan is the only terrestrial body in the outer Solar System with an atmosphere. Thicker than that of the Earth, it is composed mostly of nitrogen, but has significant amounts of methane too. On the surface of Titan, methane can exist as a liquid, and the Cassini-Huygens probe revealed possible methane lakes and rivers when it landed on the moon in 2005.

17.6.4 THE OUTER LIMITS

Moving still further out in the solar system is the region where the trans-Neptunian objects (TNOs) lie. These are rocky and icy bodies, many of which occupy the Edgeworth–Kuiper belt between about 30 and 60 astronomical units from the Sun. The most famous of these is the dwarf planet Pluto (formerly classified as a planet), along with the other named dwarf planets: Eris, Haumea, and Makemake, as well as the candidate dwarf planets Quaoar, Orcus, and Gonggong. There are probably a hundred times more TNOs than there are asteroids, although only around a thousand have been discovered to date, and those as yet unidentified likely include thousands more dwarf planets, similar in size to Pluto or larger.

Finally, the outer reaches of the Solar System are where the Öpik–Oort cloud is found. This is a spherical region stretching perhaps one-third of the way to the Alpha Centauri system (4 light-years away) and is where long-period comets originate. Comets are often referred to as "dirty snowballs" and are composed of ices and rocky particles. They occasionally stray into the inner Solar System where the heat of the Sun boils away their outer layers to produce the streaming tails that characterize their appearance in the night sky.

Water (H_2O) is not the only "icy" material that comets and TNOs contain; other simple molecules such as carbon dioxide (CO_2), ammonia (NH_3), and methane (CH_4) are also present, in solid form when the temperature is cold enough. It is speculated that comets may have played a role in transporting such molecules to the Earth, in the early years of its formation, and may even have played a role in the origin of life.

17.7   EXOPLANETS AND HOW TO FIND THEM

Thirty years ago, astronomers didn't know for certain whether there were planets in orbit around any stars other than the Sun. All that changed in 1995 with the discovery of a planet orbiting the star 51 Pegasi; today over 4000 exoplanets are known with thousands more candidates having been identified. The reasons for the huge advance are twofold: first, technology has improved to the extent that the tiny influences that a planet has on its parent star may now be measured; and second, advances in computing mean that vast quantities of data may now be automatically scanned to pick out the rare signals that indicate the presence of an exoplanet.

The problem, of course, is that planets don't generate their own light, they merely reflect the light from the star that they're orbiting around. You might think that the obvious way to look for exoplanets is simply to take very high-resolution images of stars and look for the "faint dots" nearby that are the reflected light from an orbiting planet. Unfortunately, the contrast between the star and planet limits the usefulness of this technique. The light reflected from a planet is at least millions, if not billions, of times fainter than the light emitted by a star. So the challenge is rather like looking for a mosquito lost in the glare of a streetlight, from several miles away. The difficulties can be overcome in certain cases though. If a system has a large (several times Jupiter-sized) planet, orbiting at a large distance (tens of astronomical units) away from a faint star (such as a brown dwarf), then observing in the infrared, the contrast ratio between the star and planet may be only a few tens of thousands to one. A dozen or so exoplanets have indeed been imaged in such circumstances, but this is in general *not* the way that planets around other stars are discovered.

As mentioned in Chapter 14, another fortuitous way in which exoplanets may occasionally be detected is via the technique of gravitational microlensing, when one star-system passes in front of another, so gravitationally magnifying its light. However, although Earth-mass planets can in principle be detected by this technique, nothing can be learnt about them apart from the mass, and furthermore the measurements are not repeatable. In practice therefore there are two main ways in which exoplanets are discovered and investigated, which we consider next.

17.7.1 THE RADIAL VELOCITY TECHNIQUE

In 1995, Swiss astronomers Michel Mayor and Didier Queloz were measuring spectra of stars very precisely. They were looking for spectral lines that exhibited small shifts with time that would indicate that the star was being tugged to-and-fro by an (unseen) planet orbiting around it. It is common to think that planets orbit (fixed) stars, but in fact stars and planets orbit their common centre of mass. Since stars are so much more massive than planets, the centre of mass of a planetary system is usually close to the centre of the star, so the star will not move much in response to the gravity of any planets orbiting it. However, it *will* move slightly, tracing out a tiny orbit around the centre of mass of the system, with the same period as the planet. As the star moves towards the telescope (when the planet moves away), so the lines in its spectrum will be Doppler shifted slightly towards shorter wavelengths, known as a blueshift. Then, as the star moves away from the telescope (when the planet moves in the opposite direction), so the lines in its spectrum will be Doppler shifted slightly towards longer wavelengths, known as a redshift. The cycle will repeat over the course of the planetary orbit, and the Doppler shift can be directly converted into a so-called radial velocity, which is the velocity of the star towards or away from us. By measuring the amplitude of the radial velocity curve, and the period with which the cycle repeats, it is possible to determine the ratio of the mass of the star to the mass of the planet, as well as the length of the planetary "year." Now, the spectrum of a star also allows astronomers to make good estimates of its mass and radius, by comparing the spectrum to those of other stars whose masses and radii have been determined by other means. As a result, the mass of the planet that's causing the Doppler shifts can be calculated, as well as its distance away from its host star.

What Mayor and Queloz found in 1995 was quite unexpected. The planet orbiting the star 51 Pegasi, now known as 51 Pegasi b, took only 4.23 (Earth) days to complete one orbit. It was only 0.05 astronomical units away from the star and had a mass around half that of Jupiter. Orbiting so close to its host star, it was what has become known as a hot Jupiter exoplanet. Such a planet could not have formed in its present location – it must have condensed beyond the snow line and subsequently migrated inwards towards the star to reach its current position. Since many hot Jupiters have now been found, such migration must be common, and perhaps raises the question of why Jupiter, in the Solar System, did not do likewise.

The radial velocity method is the means by which the masses of most exoplanets have now been measured. Almost 1000 planets have been discovered using this technique, and the masses of many more found by other techniques have subsequently been measured using it. The radial velocity amplitude of 51 Pegasi measured by Mayor and Queloz was only around 50 m s^{-1}, which arose from Doppler shifts of no more than one ten-thousandth of a nanometre.

Nonetheless, the technology of spectrographs is now such that radial velocities of less than one metre per second are within reach of measurement, corresponding to planets whose masses are comparable to the mass of the Earth. The very nearest star to the Sun, known as Proxima Centauri, has been discovered to host an exoplanet. Its planet, Proxima Centauri b, is only a little more massive than the Earth, but orbits its red dwarf host star in only 11 days. The resulting radial velocity amplitude is measured to be about 1.4 m s^{-1}. Similarly, the nearby Barnard's star is another red dwarf which is orbited by a 3 Earth-mass planet in a 233-day orbit causing a 1.2 m s^{-1} radial velocity amplitude.

One point that has been glossed over until now is that the radial velocity method alone does not allow an unambiguous measurement of the exoplanet mass to be determined. The problem is that without other information, it is not possible to know the inclination of the planetary orbit with respect to the line of sight. If it is assumed that the orbit is viewed edge-on, then the mass calculated from the measured Doppler shift is indeed the mass of the planet. However, a more massive planet in an inclined orbit would produce the *same* radial velocity amplitude. Since the inclination is not (in general) known, all that is measured directly is a lower limit to the planet's mass. Of course, by measuring enough exoplanets in this way, some of their orbits will be close to edge-on and others close to face-on, so on average it is possible to get a good idea of the overall distribution of planetary masses, even if it is not possible to be precise in any specific case.

Another limitation of the technique is that it's necessary to observe individual stars over the course of several nights, weeks, or months, in order to plot out the complete radial velocity curve. Individual stars must be targeted, and large amounts of time on large telescopes must be invested in order to achieve a single exoplanet mass measurement.

17.7.2 THE TRANSIT TECHNIQUE

There is a technique, however, which allows astronomers to search for planets around millions of stars simultaneously, which also allows the inclination of the planetary orbits to be determined, and which allows other information about the planet to be inferred such as its density and even its atmospheric composition. The technique is referred to as transit photometry. It relies on the fact that, if an exoplanetary orbit is observed close to edge-on, then the planet will pass across the face of its star once every orbit. When it does so, it will block out a tiny fraction of the star's light, and so the star will appear to dim slightly, once per orbit.

To pursue this technique in practice, arrays of small ground-based telescopes are used to monitor large patches of sky, taking images every few minutes over the course of years. Small ground-based telescopes with wide fields of view, such as the Wide Angle Search for Planets (*SuperWASP*) project, can image an area of sky containing perhaps a hundred thousand stars in one snapshot. Sophisticated computer programs are then used to automatically measure the brightness of every star in each image, and string together the individual measurements of each star over many thousands of images, to generate what is known as a light curve for each star. Further programs then scan through these light curves searching for the tiny, periodic dips that indicate the presence of a transiting exoplanet. A Jupiter-sized planet transiting across the face of a Sun-like star will give a dip in brightness of around 1%; for an Earth-sized planet, the dip in brightness would be 100 times smaller still at around 0.01%.

Many of the exoplanets first discovered by ground-based surveys such as *SuperWASP* are further examples of the hot Jupiter type. Some of these are in very extreme environments. For example, the planet WASP-12b orbits its host star in just over one day. It is so close to the star that its atmosphere has been inflated by the star's radiation and is boiling away into space. The planet's surface temperature is around 2500 K, and it is distorted into an egg-shape by the star's gravity.

The most successful project to detect transiting exoplanets in recent years has been the NASA *Kepler* observatory, launched in 2009. This satellite stared continuously at a patch of sky between the constellations of Cygnus and Lyra, regularly monitoring the brightnesses of each of the stars it saw. As a space-based telescope, it is sensitive enough to detect the very small dips produced by transiting Earth-sized planets. Consequently, it has found hundreds of transiting exoplanets, including many in multiple-planet systems, along with several thousand candidates.

Deciding what is a genuine transiting exoplanet, rather than something which merely mimics a transit, is a difficult business. Eclipsing binary stars in which one of the stars just grazes the limb of the other as it passes in front, can produce 1% dips too, but these dips will be V-shaped not U-shaped as those resulting from planetary transits are. Accurate photometry, such as *Kepler* obtains, can generally allow these mimics to be distinguished. Somewhat harder to rule out are eclipsing binaries whose light is diluted by another nearby star. Within a few pixels of the *Kepler* detector, it may happen that an eclipsing binary star lines up with a brighter, single star. The light from the single star will "dilute" that from the eclipsing binary, resulting in (say) a 20% deep eclipse appearing like a 0.1% depth transit, and so mimicking the appearance of a transiting exoplanet.

Even harder to rule out based on the light curve alone are transits by planet-sized stars, such as white dwarfs. The only way to rule these out is to do radial velocity spectroscopy and measure the mass of the transiting object – white dwarfs are much more massive than planets, and so those mimics can be ruled out too, if the radial velocity can be measured. The problem with the *Kepler* candidates is that most of the stars it observes are too faint to obtain detailed radial velocity spectroscopy.

17.7.3 EXOPLANETARY SYSTEMS

Where *Kepler* candidates have been confirmed as exoplanets, it is often because they are in systems with multiple planets. In those cases, the gravitational pull of each planet on each of the others can speed up or slow down their orbits, and so cause measurable delays in the times of transit. By accurately modelling the observed transit timing variations (TTVs), the masses of each of the transiting objects can be calculated, allowing them to be confirmed as genuine planets. Careful analysis of the *Kepler* data has led astronomers to conclude that almost all of its candidates in those systems appearing to host multiple planets are indeed genuine planets.

An example is the Kepler 11 planetary system of six transiting planets (named b–g). The star itself is very similar to the Sun, and the six planets all have sizes measured from their transit depths of around two to three times that of the Earth. However, the entire system is very compact, with planetary orbital periods of between only 10 and 120 days, such that the whole system would fit within the orbit of Venus in our Solar System. As a result of the observed TTVs, planetary masses of between two and ten times that of the Earth have been calculated for the six planets. Another multi-planet system discovered through TTVs, this time from ground-based observations, is Trappist-1 (Figure 17.15). This red dwarf star has seven transiting planets, all of which are comparable in size and mass to the Earth (or smaller), with orbital periods from about 1.5 days to 18 days. The entire planetary system is only around 0.1 astronomical units across.

Other interesting transiting exoplanets have been discovered by *Kepler* in binary star systems, orbiting a pair of stars. Such circumbinary planets would have the remarkable phenomena of seeing two "suns" in the sky, just like the fictional planet Tatooine in *Star Wars*. Finally, there are also now hints of exomoons from transit photometry studies too. The Jupiter-sized planet Kepler 1625b may be orbited by a Neptune-sized moon which causes an additional small dip in the transit light curve and gives rise to transit timing variations for the planetary transits too.

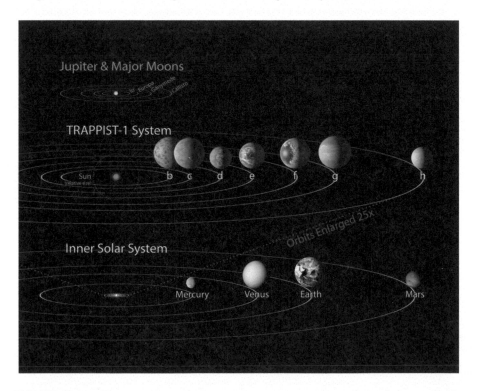

FIGURE 17.15 The Trappist-1 planetary system compared to the Solar System. (Credit: NASA/JPL-Caltech.)

Over 3000 exoplanets have currently been confirmed by the transit method. Many are hot Jupiter or regular Jupiter-type planets, but increasingly a number of mini-Neptune sized, and super-Earth sized planets, have now been found as well (such as those orbiting Kepler 11 and Trappist-1). Some of the planets discovered are in the star's habitable zone – the region of space around the star where the temperature is such that water might exist in liquid form on the surface of a planet that lies there. Alternatively, there may exist habitable exomoons orbiting planets in the star's habitable zone (Figure 17.16), perhaps reminiscent of the Ewok's home, the fictional forest moon of Endor in *Star Wars*! Clearly such moons or planets are of great interest as offering the potential for harbouring life. Discovering whether transiting exoplanets in a star's habitable zone might actually host alien biospheres is (amazingly) potentially within reach of observations.

As a first step, the depth of the transit provides a direct measure of the radius of the exoplanet, and the duration of the transit may be used to calculate the inclination of the orbit (which will always be close to edge-on, otherwise no transit would be seen). If the host star of a transiting exoplanet is subsequently observed spectroscopically, in order to measure its radial velocity curve, then the mass of the planet can be unambiguously measured too. Having determined both the mass and radius of the exoplanet, its mean density may be simply calculated, and this in turn provides an indication of the bulk composition of the exoplanet – whether it is rocky or gaseous for example.

However, the value of transiting exoplanets does not end there. During the transit itself, a tiny proportion of the star's light will reach a telescope after passing through the atmosphere of the planet. By carefully measuring the difference in the spectrum during the transit compared with the out-of-transit spectrum, the atmospheric

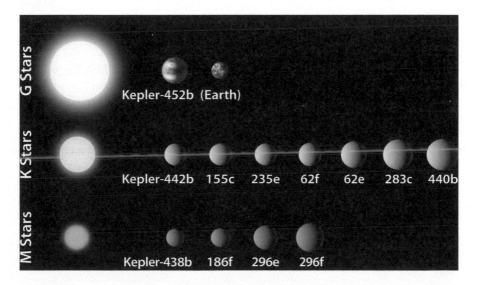

FIGURE 17.16 Habitable zone exoplanets discovered by *Kepler*. (Credit: NASA Ames/JPL-CalTech/R. Hurt.)

composition of the exoplanet may in principle be inferred. Ultimately, this may offer the first chance of detecting alien life elsewhere in the Galaxy. Such life may only be a green slime that you could scrape off a rock with your finger, but it would nonetheless still be life. If a transiting exoplanet hosts a biosphere, then its atmosphere may well contain oxygen, ozone, water vapour, methane, and other biosignature gases. The detection of such an atmosphere could well be the first indication that the Earth is not the only inhabited planet in the Galaxy.

17.8   LIFE IN THE UNIVERSE

Statistical analyses of the results of exoplanet surveys are now reaching the conclusion that planets are ubiquitous in the Galaxy: stars are orbited by planets as a rule, rather than the exception. The question naturally arises as to whether any of these planets (other than the Earth) might harbour life?

17.8.1   THE DRAKE EQUATION

In 1961, whilst preparing to host a meeting about how to detect extra-terrestrial intelligence, radio astronomer Frank Drake simply wrote down all the things that astronomers needed to know in order to predict the answer to the question: what are the chances of detecting extra-terrestrial civilizations, via their radio communications? He realized that if all these quantities were multiplied together, they gave the number of detectable civilizations in the Galaxy. His equation can be written as follows:

$$N = R^* \times f_p \times n_e \times f_l \times f_i \times f_c \times L$$

The seven terms encapsulate, in a simple way, all the parameters of stars, planetary systems, and life that combine together to give the number of detectable civilizations. Some of these terms are known pretty well; for others there is virtually no idea.

Firstly, R^* is the average star formation rate in the Galaxy, which is known to be of order ten per year and is perhaps the parameter that is known most precisely.

The number f_p is the fraction of all stars that have planets. As a result of recent exoplanet surveys, it now seems likely that this number is close to 1 and is at least 0.5. The parameter n_e is the average number of planets, per star that has planets, that can support life. In the Solar System, the number of planets in the habitable zone is clearly at least one (the Earth). However, it's plausible that Mars too may have supported primitive life in the past, and even Venus may not be ruled out as an abode of life, as hinted at by the recent detection of phosphine in its atmosphere, which may have a biogenic origin. For this reason, adopting a value of two for this parameter is not unreasonable.

The remaining parameters are far less certain. f_l is the fraction of planets that actually develop life at some point; f_i is the fraction of those that develop intelligent life; and f_c is the fraction of those that develop a technological civilization which is capable of communicating its presence to others in the Galaxy. Estimates for each of these parameters range from "negligible" ($f = 0$) to "certain" ($f = 1$) in each case!

Finally, L is the length of time over which a civilization maintains the ability to communicate over interstellar distances. This could perhaps be as short as 100 years but may be as much as a million years or more.

Frank Drake himself assumed that *all* planets that *can* support life, *will* develop it, and further that about 1% of those would develop into intelligent life, and 1% of those would give rise to communicating civilizations which lasted around ten thousand years. His values therefore gave $N = 10 \times 0.5 \times 2 \times 1 \times 0.01 \times 0.01 \times 10{,}000 = 10$ civilizations that are detectable at any instant in the Galaxy. Plausible estimates using alternative numbers might range from $N = 0$ to $N = 20$ million!

Remember though, that this is only the number of *communicating*, intelligent, civilizations. If the concern is only with detecting biospheres on transiting exoplanets that may be due to green slime, the f_i and f_c terms are irrelevant, and L may be billions of years. In that case, the number of detectable biospheres may in principle be of order billions.

17.8.2 THE FERMI PARADOX

If the number of communicating civilizations in the Galaxy really is greater than one, as some estimates of the Drake equation would have it, the question then arises – where is everybody? The apparent paradox between the predicted number of civilizations in the Galaxy, and the lack of any evidence for them, was expounded by the physicist Enrico Fermi in 1950.

If there were even *one* other alien civilization in the Galaxy, the chances are that it would be far in advance of human technological capability. The Galaxy has been around for 10 billion years, but human-kind has only been technologically capable of communicating into space for less than 100 years. The chance that any other civilization is at precisely the same point in its history as humans are is vanishingly small; it is vastly more likely that it reached the same stage millions, or even billions, of years ago. Such a civilization could have colonized the entire Galaxy within a few million years – if not "in person," then possibly using self-replicating von Neumann machines (i.e. spacecraft that travel through space, building copies of themselves when they encounter resources to do so). Even travelling at a speed comprehensible today, such as that of the deep space probe Voyager 1 which is currently travelling out of the Solar System at around 17 km s^{-1}, it is possible to cross the Galaxy in less than two billion years. Travelling a hundred times faster (surely within the capability of an "advanced" civilization?), the journey time reduces to a mere 20 million years. So even if humans were not actively looking for alien life – if it exists, it should have found the Earth and be here already.

There are, of course, many possible solutions to the Fermi paradox, some of which are considered here. The first three possibilities come down to perhaps the most obvious answer, namely that humans are indeed alone.

1. Rare Earth – it may be that there are no civilizations in the Galaxy any more advanced than humans are. Perhaps the combination of astronomical, geological, chemical, and biological factors needed to allow the emergence

of complex, multicellular life is just so unlikely, that it's only happened once, and the parameter f_l in the Drake equation is close to zero. Clearly for life on Earth to have arisen, it was necessary to have just the right mix of chemicals present on Earth, with the Earth at just the right distance from the Sun to be at an acceptable temperature. Other factors such as the presence of a large Moon to stimulate tides, the presence of Jupiter to deflect incoming asteroids, and the existence of plate tectonics, may each contribute to the suitability, or not, of a planet to host life.

2. Doomsday – perhaps life emerges often, and civilizations also commonly arise, but it is the nature of "intelligent" life to destroy itself within a few hundred years, and so the parameter L in the Drake equation is very small. The human race certainly has no shortage of ways of accomplishing this, whether it's via physical, chemical, or biological weapons of mass destruction, or as a result of climate change, or even a nanotechnology catastrophe that renders the whole planet into a "grey goo." If life doesn't persist very long on any planet, it should not be expected to leave any evidence of its presence around the Galaxy.

3. Extinction – even if humans don't wipe themselves out, perhaps the Universe conspires to eliminate civilizations on a regular basis? It's clear on Earth that there have been at least five mass extinctions of virtually all life in the distant past. Some of these may have been triggered by the impact of massive asteroids, but other possible extinction-causing events might include nearby supernovae or gamma-ray bursts. Such considerations might also lead to small values of L in the Drake equation.

There is another class of possible solutions to the Fermi paradox that boil down to the fact that alien civilizations *do* exist, but no evidence of them may be found.

4. Distance scales – perhaps civilizations are spread too thinly throughout the Galaxy to effectively communicate with each other? Civilizations may be separated in space, and also in time, so two civilizations just don't overlap during the time that they're each active. Even if it's physically possible to spread throughout the Galaxy on a relatively short timescale, maybe it's just too expensive for a civilization to do so, requiring more resources than can be made available.

5. Technical problems – maybe humans are not looking in the right place, or in the right way? Or maybe the searches just haven't been going for long enough? Perhaps a signal that's out there has not been recognized, because the alien civilization is using technology that simply cannot be comprehended by humans. Maybe they are too alien, or too technologically advanced for humans to communicate with.

6. Isolationist – perhaps the aliens *are* out there, but they're choosing to hide themselves from Earth? Perhaps everyone is listening, but nobody is transmitting? It may be that other civilizations know the Earth is here, but it is purposely isolated, as if humans were some kind of exhibit in a zoo.

Finally, there are of course the more extreme possibilities, such as that the Galaxy which is observed to be empty of life is just a simulation, constructed by aliens, and designed to appear that way to humans. Or perhaps the aliens *are* already here, or at least were in the past, and the fact has not been recognized. Such speculation is great for science fiction, but without evidence, it's not worth pursuing such ideas further in a book such as this.

17.8.3 THE ANTHROPIC PRINCIPLE

The one place we can be sure that life exists in the Universe is here on Earth. And the very fact that we, humans, are here to observe the Universe and attempt to understand it already places strong constraints on its existence. These ideas were first encapsulated by American physicist Brandon Carter in 1973 when he introduced the term "anthropic principle" and later expanded by the cosmologists John Barrow and Frank Tipler in their book, *The Anthropic Cosmological Principle*, published in 1986.

In its simplest terms, the weak anthropic principle states that if the physical laws of the Universe were *not* such that galaxies, stars, the Sun, Earth, and life could exist, then we (or any other observer) would not be here to discuss it.

To take one example – you've already seen (in Chapter 13) that if W bosons were less massive than they actually are, the weak nuclear interaction would be stronger, beta-decays could happen more readily, and free neutrons would turn into protons much more readily. As a result, there would have been no elements other than hydrogen in the early Universe. At the other extreme, if W bosons were even more massive, weak interactions would be weaker still, free neutrons would be stable, and the entire primordial content of the Universe would be composed of helium. The fine-tuning of the W boson mass necessary to produce the Universe we inhabit is just one instance of how the Universe is apparently finely tuned to our existence.

Another, more esoteric, example concerns the dimensions of space and time. It has been known for a century that if there is only one dimension of time but more than three spatial dimensions, then the orbit of a planet around a star cannot be stable. Maxwell's theory of electromagnetism also only works if there are three spatial dimensions and one dimension of time. Furthermore, it has been shown that if there were more than one dimension of time, then protons and electrons would decay into particles heavier than themselves, and if there were fewer than three dimensions of space then the Universe would be too simple to allow complexity and observers to develop.

A final example of fine-tuning concerns the synthesis of carbon-12 via the triple-alpha process described earlier. It's hugely unlikely that three helium-4 nuclei would fuse simultaneously to make a carbon nucleus in one step. Instead what happens is that two helium nuclei fuse to first make an intermediate nucleus of beryllium-8. As noted in the discussion of the early Universe, this nucleus is unstable and will decay with a half-life of one-tenth of a femtosecond (10^{-16} s) into two helium-4 nuclei. However, its lifetime is *just* long enough that it can sometimes fuse with a third helium-4 nucleus to make carbon-12. This is not the only lucky break though. It turns out that the ground state of carbon-12 has an energy that is about 7.3 MeV *below* that of the beryllium-8 plus helium-4, so it would normally not be expected to form. Luckily there is an excited state of carbon-12 with an energy of about 0.3 MeV *above* that of the combined beryllium-8 and helium-4. This allows the two nuclei to use their kinetic energy to fuse and form the excited carbon-12 nucleus, which subsequently makes a transition to its stable ground state. According to a calculation published in 1989, the energy level of this excited state of carbon-12 must be finely tuned to between 7.596 MeV and 7.716 MeV above the ground state, in order to produce the amount of carbon-12 that we see in the Universe, and so permit life to exist.

The physical laws of the Universe are indeed finely tuned to allow our existence, but the weak anthropic principle may be nothing more than a tautology – things are the way they are because if they weren't, we wouldn't be here to discuss them. At the very least it is the result of a selection bias, implying that out of many possible universes, ours is one that allows us to exist.

However, there is also what is known as the strong anthropic principle which goes somewhat further. According to Barrow and Tipler, this states that "the Universe must have those properties which allow life to develop within it at some stage in its history." It is difficult, if not impossible, to test the truth of this principle, and it remains controversial.

Before leaving this discussion, I should also note that Tipler goes even further and proposes a final anthropic principle, which says "intelligent information-processing must come into existence in the Universe, and, once it comes into existence, will never die out." This statement is based on the idea that, for the Universe to physically exist, it must contain observers to perceive it. Since the Universe obviously does exist, it must sustain life forever. The idea of observers being necessary for something to exist arises from some interpretations of quantum theory, but here we are perhaps entering the realm of philosophy rather than physics.

Instead, the Universe as it appears today is now left behind as the next chapter considers what the future holds in store.

18 The Future of the Universe

So what is the future of the Universe? Will it go on expanding and cooling forever? Or does it have another fate in store? Amazingly, cosmologists are now in a position to answer this ultimate question with a fair degree of certainty.

Before doing so, a final piece of terminology should be explained at this point. Just as we use the capitalized word "Galaxy" to refer to our own Milky Way and "galaxy" (un-capitalized) to refer to other similar structures, so we use the capitalized word "Universe" to refer to the entirety of all matter, energy, and spacetime that exists, whereas "universe" (un-capitalized) is used to refer to certain alternative scenarios that we consider in this chapter.

In order to explore conclusions about the exact fate of the Universe and its alternatives, it is first necessary to consider one more large-scale property of space – namely its geometry.

18.1   THE GEOMETRY OF SPACE

On an everyday scale, most people are familiar with the idea that parallel lines remain parallel, no matter how far they are extended in any direction, and that the three angles inside a triangle add up to 180°. These two geometric examples correspond to what may be called flat space or space with zero curvature. However, in the curved four-dimensional spacetime near massive objects, these two rules may not apply! You have already read (in Chapter 14) about the way that matter curves space according to Einstein's general theory of relativity, so it should come as no surprise that the geometry of space depends on the amount of matter within it.

In a space with positive curvature, initially parallel lines eventually converge, and the angles inside a triangle add up to more than 180°. A familiar two-dimensional example of such a space is the surface of a sphere. Conversely in a space with a negative curvature, lines that are initially parallel eventually diverge, and the angles inside a triangle add up to less than 180°. A less familiar two-dimensional example of this type of space is the surface of a saddle. Picturing the equivalent geometries for four-dimensional spacetime is not easy to do, but the mathematics of such spaces can be extrapolated from the more familiar examples that exist in fewer dimensions (Figure 18.1).

The curvature of space is determined by the overall amount of mass and energy in the Universe. (Remember that mass and energy are equivalent via $E = mc^2$.) In particular, it is useful to refer to the density of the Universe, where both mass and energy can contribute to the overall density. A universe with a *high* density can have a *positive* curvature, whereas a universe with a *low* density can have a *negative*

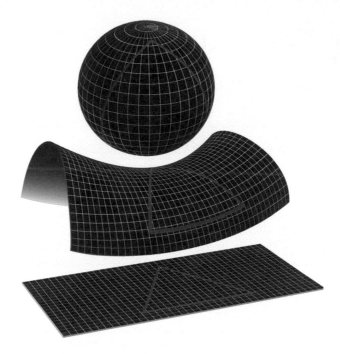

FIGURE 18.1 The curvature of space: positively curved, negatively curved, and flat two-dimensional surfaces. (Credit: NASA/WMAP science team.)

curvature. The dividing line between the two is the critical density, which corresponds to zero curvature.

Needless to say, it is very difficult to measure the large-scale curvature of space, because any effect is negligible on the relatively small, everyday scales over which it might be hoped to measure it. Indeed, on small scales, the geometry of the Universe appears to correspond to flat space or zero curvature. Remarkably, a crucial result of Alan Guth's inflation theory is that the very rapid expansion in the first fraction of a second drives the Universe on the very largest scales also to have zero curvature. Any deviations from the critical density that may have existed before the period of inflation are smoothed out by the inflationary process, resulting in a universe with almost *exactly* the critical density, to a precision of better than one part in a million. In other words, if the inflation theory is correct, the Universe is indeed one of flat space.

The overall density of a universe whose geometry is flat must be precisely equal to the critical density. If the Hubble constant is currently about 70 km s^{-1} Mpc^{-1}, then the critical density is currently equivalent to about four protons per cubic metre. If the average density were to be less than the critical value, such a universe would have a negative curvature; if the average density were to be greater than the critical value, such a universe would have a positive curvature. Perhaps not surprisingly, it therefore turns out that the ultimate fate of the Universe depends crucially on its density.

18.2   OPEN AND CLOSED UNIVERSES

The Universe is currently expanding and, broadly speaking, the two options for its future behaviour are that it may expand forever or it may not. To decide which it will do, it is instructive to turn for help to an analogy, namely that of launching an object from the surface of the Earth. If a rocket is launched with a small amount of kinetic energy, it will eventually fall back to Earth as gravity wins and pulls the rocket back. If the rocket is given a larger amount of kinetic energy, sufficient for it to exceed the escape speed required to leave the Earth completely, the rocket will do so and travel out into space.

As it is with the rocket, so it is with the Universe. The balancing act here is between the energy of the expansion of the Universe (the kinetic energy) and the gravitational potential energy of all the matter in the Universe. If the energy of the expansion is sufficient to overcome the gravitational pull, then the Universe will continue expanding forever – such a universe is known as "open." If, on the other hand, the energy of expansion is not enough to overcome the gravitational pull, the expansion will gradually slow down, and ultimately reverse – such a universe is known as "closed."

If the overall density of the Universe is greater than the critical density, space has a positive curvature, and such a universe is closed. A closed universe is one in which gravity wins; it is *finite* in size at all times. The expansion that is currently observed would gradually slow down and eventually stop altogether. But gravity doesn't give up there! The expansion would then *reverse* – all the matter in such a universe would begin to converge as space contracts. The contraction would gradually speed up and all the galaxies and clusters would rush towards each other.

During the contraction, all the processes of the Big Bang outlined in Chapter 16 would occur in reverse order! The atoms would ionize under the impact of radiation; then the nuclei would be smashed apart into protons and neutrons; finally, the nucleons themselves would disintegrate into their constituent quarks. Photons would spontaneously create pairs of particles and antiparticles until equal amounts of radiation and matter again filled such a universe. As the temperature of the universe rose, so the four interactions would each in turn become indistinguishable as the unifications proceeded in reverse order.

Ultimately things would reach a mirror image of the Big Bang – known as a big crunch. And what would happen next? Well, one possibility is that's the end of everything – no more matter, no more space, no more time. Such a universe just ends.

But maybe not… Some cosmologists suggest that what happens instead is a kind of big bounce. At the very last instant, the whole sequence would turn around and a new Big Bang would happen. There would be a whole new period of expansion and contraction, repeating the history of such a universe again and again, over and over, forevermore. What is known as *the* Big Bang may simply be the latest in an infinite series of big bounces that is set to repeat an infinite number of times. The very latest ideas on this topic are revisited at the end of the book.

Returning to the second possibility, if the overall density of the Universe is less than the critical density, space has a negative curvature, and such a universe is open.

An open universe is one in which gravity loses, so that the separation between objects would continue to increase forever; it would be *infinite* in size at all times. It would have been infinite in size at the instant of the Big Bang and would remain infinite in size as space expands and the separation between galaxies increases. In the future of such an ever-expanding universe, all the stars would eventually run out of nuclear fuel. Low-mass stars (such as the Sun) will evolve into white dwarfs; more massive stars will undergo supernova explosions, before ending up as either neutron stars or black holes. These dead stars, along with any planets and other pieces of rock and dust, would gradually spiral in towards the centres of their respective galaxies where they would be consumed by the massive black holes that exist in most galaxies. So, at this point such a universe would be cold and dead, containing nothing but black holes and cosmic background photons, but still continuing to expand.

You may think that this is the end of the story – but not so. The name black hole was given to these objects in the belief that nothing – not even light – can escape from them, so they would literally appear to be a black hole in space. But the British cosmologist Stephen Hawking showed that black holes are not entirely black. By a neat trick of quantum physics, black holes will eventually "evaporate" into a swarm of subatomic particles and antiparticles. The process, however, takes an extremely long time – a black hole with the mass of a whole galaxy will take about 10^{97} years to evaporate! But that's okay, because such a universe would have an infinite time to expand in this scenario. These particles and antiparticles would eventually mutually annihilate each other, each pair creating photons of electromagnetic radiation. So, the *final* fate of an open universe is that it contains just photons, and simply becomes more and more dilute as the expansion continues. This is known as the heat death of a universe.

This discussion of open and closed universes is all very well, but it has already been noted that the Universe apparently has a geometry corresponding to flat space, and so must have an overall density that is exactly equal to the critical density. This places it precisely on the dividing line between the open and closed universe models.

In fact there is also a third possibility for the fate of the Universe – called by cosmologist John Barrow the "British compromise universe." This is a universe that is not quite open, but not quite closed either. After all, if there are these two extremes – a closed universe in which gravity wins and an open universe in which expansion wins – there must be a critical situation in between the two in which the energy of expansion is *exactly* balanced by the gravitational pull. One proton more and gravity would win, leading to a big crunch; one proton fewer in this universe and the expansion would win, leading to the heat death scenario. In this model, more properly called the critical universe model, the entire universe is infinite in size at all times, just as in the open universe model. However, it has the interesting consequence that although the rate of expansion continues to slow down, it only reaches zero expansion rate at an infinite time in the future, at which point all objects in such a universe have an infinite separation. A universe like this will also suffer a heat death just like an open universe.

18.3 A CLUE FROM SUPERNOVAE

From the preceding discussion you might conclude that the consensus amongst cosmologists is that the expansion of the Universe has continuously slowed down (decelerated) since the time of inflation, and the Universe is now following the expansion rate described by the critical universe model. That is to say, the Universe will expand forever, but at an ever-decreasing rate, such that objects will reach an infinite separation at an infinite time in the future.

Well, that *was* the accepted picture of the Universe until 1998. In that year, evidence began to appear that the expansion rate of the Universe may not be as straightforward as astronomers had previously believed. The new observational evidence was based around measurements made of supernovae in distant galaxies.

As you know, supernovae are the result of exploding stars. The sort previously described in Chapter 17, known as type II supernovae, are believed to be caused by massive stars that reach the ends of their lives and then explode when they cannot undergo any more energy-releasing fusion reactions in their cores. But another sort, mentioned in Chapter 9, type Ia supernovae, are something quite different. Their spectra show no evidence of hydrogen and they all seem to reach the *same* luminosity. And this is where their usefulness as a standard candle and a probe of the distant Universe comes in. If they all have the same luminosity, and astronomers know what that luminosity is, and if their apparent brightness (or flux) can be measured, then astronomers can calculate how far away they are. Furthermore, by looking at the relative positions of lines in their spectra, astronomers can also measure their redshift. So in principle, astronomers can measure both how far away type Ia supernovae and their host galaxies are *and* the apparent speed with which their host galaxies are receding from us.

In 1998, astronomers from the "Supernova Cosmology Project," led by Saul Perlmutter of the Lawrence Berkeley Laboratory in California, reported measurements of 42 distant type Ia supernovae. At around the same time, another team of astronomers from the "High z Supernova Search Team" led by Adam Riess of the Space Telescope Science Institute and Brian Schmidt of Mount Stromlo Siding Springs Observatories, presented measurements of a further 16 distant type Ia supernovae. Both sets of supernovae had redshifts between about 0.3 and 0.8, so light from them was emitted between about 3 and 7 billion years ago. When the maximum brightnesses of these supernovae were plotted against their redshift, the results from both teams revealed the same thing – that the distant supernovae are about 15–25% *fainter* than they ought to be if the Universe had simply followed the expansion rate predicted by a conventional critical universe model. The proposed explanation for why these supernovae appear fainter than expected is that the light from the supernova explosions has travelled a greater distance than is predicted by the conventional model. The researchers suggest that the reason for this is that the Universe's expansion *speeded up* while the light from the supernovae was on its way to the Earth, so the amount of expansion has been greater than predicted by the conventional model. In other words, the expansion of the Universe is currently speeding up (accelerating), not slowing down (decelerating), as everyone had assumed until recently.

Astronomers realized that this suggests the presence of a mysterious form of "negative gravity" or so-called dark energy, which *opposes* the gravitational attraction of matter that was assumed to be responsible for the deceleration of the universal expansion. Dark energy seems to be a property of empty space – even space with no matter particles in it – and it behaves rather differently from the matter or radiation in the Universe. As space expands, the density of matter and radiation must both get progressively less and less as things become further and further apart. However, the dark energy of empty space can remain *undiluted* by this expansion and so exert a significant effect on the expansion rate of the Universe.

In fact, "negative gravity" is not a new idea. Albert Einstein himself first suggested the idea when he developed his general theory of relativity in 1915. Having derived the equations that describe the overall behaviour of the Universe, Einstein was concerned that his equations did not allow a static universe to exist. According to the equations of general relativity, the Universe had to be either expanding or contracting. At that time there was no evidence that the Universe was anything but static, so Einstein introduced an extra term into his equations which he called Λ (the Greek upper-case lambda) – the so-called cosmological constant. This took the form of a sort of negative gravity and was assumed to be just strong enough to counteract the pull of gravity and so result in a static universe. When, a few years later, Edwin Hubble announced his evidence that the Universe is in fact expanding and not static at all, Einstein is said to have commented that the introduction of the cosmological constant was the biggest blunder of his life. However, in the light of the type Ia supernova results, it seems Einstein was on the right track after all…

Before getting carried away with this idea though, it is worth looking at these supernova results more closely. First, the values for the brightnesses of the distant supernovae are rather uncertain, and it is still just about conceivable that the conventional expansion rate would fit the data. Secondly, there is the calibration problem. Astronomers need to know the maximum apparent brightness of the supernova, but comparing supernovae at different redshifts means that different parts of the supernova spectrum are shifted into or out of the spectral range that is used to measure the brightness (say the optical range between about 400 nm and 700 nm). This effect must be corrected for in order to work out the actual maximum brightness in a comparable way for all type Ia supernovae whatever their redshift, and clearly there is the potential for uncertainties in the calibration process. Thirdly, astronomers must also take account of the amount of absorption that the light from the supernovae has experienced as it travels huge distances across the Universe. More absorption, caused say by extra dust in the vicinity of the supernova itself, could also make the distant supernovae appear fainter than expected. Even if these problems are solved though, there is still one more difficulty. Just what *are* these type Ia supernovae and how certain is it that they all *do* have the same luminosity?

The short answer is that no one really knows exactly what type Ia supernovae actually are. What is reasonably clear is that they are the result of white dwarfs that somehow get pushed over the maximum mass at which they can survive. As noted earlier, white dwarfs are the dead cores of low-mass stars, supported against collapse due to gravity by electron degeneracy pressure. As you know, this effect is unable to support a white dwarf against collapse if the white dwarf mass is greater than the Chandrasekhar limit. The idea is that some white dwarfs suddenly find themselves with greater than 1.4 times the mass of the Sun and so collapse, giving rise to a supernova explosion. Because all white dwarfs have the same maximum mass, this is supposed to be the reason why all type Ia supernovae have the same luminosity. But what are these objects *before* they go bang?

One idea is that they are a binary star system consisting of a pair of white dwarfs whose combined mass is greater than 1.4 times that of the Sun, such as seen in some ultracompact X-ray binary stars as mentioned earlier. As the stars orbit each other, they lose energy by radiating gravitational waves and gradually spiral together. When they get close enough, the two white dwarfs merge and a type Ia supernova is the result. The planned space-based gravitational wave detector, *LISA*, should be sensitive enough to detect gravitational waves from merging white dwarfs, and so may confirm or refute this suggestion.

Alternatively, type Ia supernovae may arise in cataclysmic variable stars, in which a white dwarf constantly pulls material from a companion star. The strong gravitational field of the white dwarf is able to tear its companion apart, and gradually more and more mass is deposited on the white dwarf surface. Some of this material undergoes nuclear fusion on the white dwarf surface and may be blown away from the system. However, more accumulates than is lost, so the mass of the white dwarf steadily increases. The white dwarf eventually reaches a mass greater than 1.4 solar masses and a type Ia supernovae is the result.

Astronomers do see examples both of binary white dwarfs and of white dwarfs pulling material off companion stars and undergoing constant nuclear fusion on their surfaces in the local neighbourhood of the Galaxy. Either, or possibly both, of these types of system may be the source of type Ia supernovae in the Galaxy and in other galaxies throughout the Universe. But will both give rise to type Ia supernovae of the *same* luminosity? And will that luminosity always be the same at all times in the Universe's history (and so at all redshifts)?

The chemical composition of the Universe has certainly changed with time as later generations of stars have formed from material that has been enriched by heavy elements formed in, and expelled by, earlier supernovae. Might the increasing metallicity mean that modern (nearby) type Ia supernovae are intrinsically *less luminous* than ancient (distant) type Ia supernovae? At the time of writing, no one knows the answers to any of these questions, although there is a great deal of current research being devoted to finding out the answers.

18.4   ACCELERATION AND DECELERATION

An indication that the acceleration and dark energy idea may be correct came in April 2001 when Adam Riess and his team were studying images of the Hubble Deep Field. These were the deepest exposure images ever made by the Hubble Space Telescope and so contained some of the faintest and most distant galaxies ever seen. By comparing two images, taken in 1995 and 1997, Riess discovered the most distant supernova then seen. It was a type Ia supernova and it had a redshift of 1.7, which means that the light from it was emitted when the Universe was only about 4 billion years old. The extraordinary thing is that the maximum brightness of this supernova was roughly twice as large as it should be! So this supernova must be significantly *closer* than it would be if the Universe had expanded at a steady rate.

The proposed explanation for this is as follows. Soon after the Big Bang, when the Universe was only a few billion years old, galaxies were relatively close together and their gravitational pull was sufficient to slow down the expansion of the Universe and so produce a deceleration. A supernova that exploded during this period, such as the one whose discovery was announced in April 2001, would thus be closer now than is suggested by its measured redshift. As the Universe aged and galaxies grew farther apart, dark energy was able to win out over the gravitational attraction, so making the Universe expand ever faster and turning the deceleration into an acceleration.

As astronomers look further back in time they should see the (current) acceleration turn into a deceleration – and this is just what the various observations of type Ia supernovae imply. The supernovae originally observed by the "Supernova Cosmology Project" and the "High z Supernova Search Team" in the redshift range 0.3–0.8 are fainter than expected (they are in the nearby, recent, accelerating Universe), while that measured by Adam Riess, at a redshift of 1.7 is brighter than expected (it is in the more distant, early, decelerating Universe).

In fact, in the last decade, the discovery of more of these distant supernovae with redshifts greater than 1.2 has confirmed the results. It now seems beyond doubt that the expansion of the Universe is currently accelerating, but that phase didn't begin until the Universe was more than half its present age. The next section explores the consequences for the Universe of these type Ia supernovae measurements and their interpretation.

18.5   MATTER AND ENERGY IN THE UNIVERSE

As mentioned earlier, inflation implies that the geometry of the Universe corresponds to flat space or zero curvature and as such its density should be equal to the critical density. Astronomers indicate this by saying that the actual density divided by the critical density is equal to one, or in symbols $\Omega = 1$, where Ω is the capital letter omega – the last letter of the Greek alphabet.

If this critical density is all accounted for by matter, then astronomers say that "omega matter" is equal to one, i.e. $\Omega_M = 1$. A flat universe in which the critical density is all accounted for by matter would follow a critical universe model as discussed earlier; it would expand forever, but the expansion would continuously decelerate, and such a universe would end in a heat death. However, the recent type Ia supernovae results imply that there is another sort of density acting in the Universe – the density of dark energy, represented by "omega lambda," Ω_Λ. Dark energy can also contribute to the density of the Universe because, according to Einstein's famous equation $E = mc^2$, mass and energy are interchangeable. It is possible to have a *flat* but *open* universe, i.e. one that expands forever at an *increasing* rate, if $\Omega_M + \Omega_\Lambda = 1$. The ultimate fate of a flat universe with dark energy present may be something quite different to the fate of a matter-only universe previously described, as you will see in the next section.

So what do the latest results imply for Ω_M and Ω_Λ? The best agreement with the type Ia supernovae data and with results from observations of the CMB is found to be a universe model with Ω_M equal to about 0.31 and Ω_Λ equal to about 0.69. Therefore the Universe does indeed have $\Omega_M + \Omega_\Lambda = 1$. But what evidence is there that the Universe contains an amount of matter equal to 31% of the critical density, implied by $\Omega_M = 0.31$?

Current understanding of nucleosynthesis in the first three minutes after the Big Bang predicts that the density of baryonic matter in the Universe (i.e. protons, neutrons, and electrons forming hydrogen, helium, and other elements) amounts to about 5% of the critical density. However, adding up the amount of matter implied by all the stars and galaxies that are seen in the Universe, astronomers measure that the fraction of the critical density accounted for by luminous matter is only about 0.5%. Even massive black holes, presumed to exist in the centres of galaxies, can only account for another 0.0001% or so of the critical density, so their contribution is negligible. This is known as the missing baryon problem.

The solution to the problem is that baryonic dark matter also exists in the Universe. As its name suggests, this matter cannot be seen even though it is also made of the familiar protons, neutrons, and electrons that stars and galaxies are made of. In fact, the idea of dark matter dates back to the early 1930s when Dutch astronomer Jan Oort and Swiss astrophysicist Fritz Zwicky independently suggested that the mass in the plane of the Galaxy and in other galaxy clusters must be greater than what was observed. Later, in 1978, American astronomers Vera Rubin and Kent Ford showed that dark matter is needed to explain the rotation curves of galaxies and concluded that galaxies must contain about six times as much mass in the form of dark matter as in the form of visible matter.

A fraction of the baryonic dark matter is almost certainly in the form of low-mass dead stars, that were never able to undergo nuclear fusion. These objects make their presence felt by the motion of stars in the outer regions of some galaxies. Searches for these massive astrophysical compact halo objects, known as MACHOs for short, in the outer regions of the Galaxy are currently underway and beginning to bear fruit. Another place where baryonic dark matter may be hiding is in diffuse, hot $(10^5-10^7$ K) strands of matter between galaxies, known as the warm-hot intergalactic medium or the WHIM for short (Figure 18.2). It is now believed that MACHOs and the WHIM together can indeed account for the predicted 5% of the critical density that exists in the form of baryons.

Now that the location of the 5% of the Universe that is in the form baryons has been settled, where is the other 26% or so of the critical matter density? Quite simply the implication is that a huge fraction of the matter in the Universe is made of some (so far unknown) constituents that are totally unlike the rest of the protons, neutrons, and electrons from which the familiar contents of the Universe are built. This is called non-baryonic dark matter. Its constituents are often referred to as weakly interacting massive particles, known as WIMPs for short, and a great deal of effort is currently being expended on searching for it!

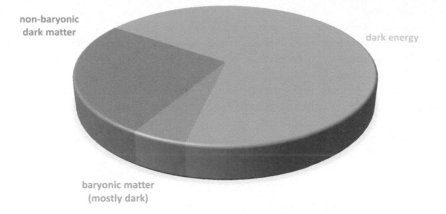

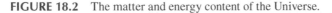

FIGURE 18.2   The matter and energy content of the Universe.

An alternative name for dark energy is quintessence. This is a word meaning literally fifth essence and follows from the ancient idea of the "four elements" of earth, air, fire, and water. In the Universe however, the five "elements" that comprise it are baryons, leptons, photons, dark matter, and dark energy. Notice that even though leptons and photons are ubiquitous in the Universe, and there are ten billion photons for every baryon or lepton, their contributions to the energy content of the Universe at the present time are negligible.

18.6   THE FATE OF THE UNIVERSE

Having arrived at an understanding that the Universe consists of 69% dark energy, 26% non-baryonic dark matter, and 5% baryonic matter (most of which is also dark), we can finally answer the question: what is the likely fate of the Universe?

We know the expansion of the Universe is currently accelerating, but it is not clear whether or not this will simply continue forevermore. It turns out that the answer depends crucially on the *type* of dark energy that exists, and specifically on the ratio between its pressure and its energy density (i.e. its energy per unit volume). This ratio is referred to as the equation of state parameter.

For a regular gas, the equation of state parameter is essentially zero. The energy density of matter (Einstein's mc^2 per unit volume) is very large, whilst the pressure it exerts is much lower, unless the temperature is extremely high. Therefore, the ratio between the two is very small in the current Universe. For electromagnetic radiation (photons) the equation of state parameter is +1/3. The consequence of this is that the energy density of radiation decreases more quickly than the volume expansion of the Universe, because its wavelength is redshifted.

For dark energy, the equation of state parameter must be less than −1/3 in order to cause an acceleration. If it is between −1/3 and −1, then the dark energy will dissipate as time progresses, so the acceleration will eventually reduce and the Universe will end in the heat death scenario that was described earlier for a matter-dominated flat universe model. However, if the equation of state parameter is less than −1, then the density of dark energy will continue to increase as the Universe expands. As a result, the accelerating expansion will proceed at an ever more rapid rate and will eventually tear the Universe apart in an event known as the big rip.

If the equation of state parameter for dark energy is, say, −1.5 then the big rip will happen in about 22 billion years' time. As this moment is approached, galaxies will first become separated from each other with about 200 million years to go. Then, with only 60 million years left, galaxies themselves will break apart as gravity becomes too weak to hold them together. In the last 100 days or so, planetary systems would also fly apart, and in the last few minutes stars and planets would disintegrate. Finally, atoms would break up in the last 10^{-19} seconds before spacetime itself is torn apart.

The crucial number that determines the ultimate fate of the Universe is therefore the equation of state parameter for dark energy. The evidence we currently have for this shows that it is indeed very close to −1, but it is not measured precisely enough to distinguish whether it is greater than, less than, or exactly equal to this value. If it is slightly less than −1, then the closer it is to that value, the further in the future the big rip lies. If it is slightly more than −1, then the big rip will never occur, and if it is exactly equal to −1, then the big rip will occur at an infinite time in the future.

18.7   BORN OUT OF FIRE: AN ALTERNATIVE TO THE BIG BANG

Before finishing the exploration of the Universe, it is instructive to mention a current challenger to the title of the theory for the origin of the Universe. The material presented in this section is absolutely at the forefront of current scientific understanding. It is also even more bizarre than anything presented so far in this book! So sit back and prepare for a final roller coaster ride to the frontiers of modern physics.

The ekpyrotic universe is the name given to a remarkable theory that provides an interesting alternative to the standard inflationary, hot Big Bang model that has permeated the last two chapters of this book. "Ekpyrotic" is a Greek word that comes from an ancient idea that the Universe is continually destroyed and then recreated by fire in an endless cycle. The new theory provides an explanation for the origin of the hot Big Bang and does away with the need for a period of inflation. It is also intricately linked with the ideas in M-theory about superunification discussed in Chapter 15 and provides perhaps the ultimate link between particle physics and cosmology. In this spirit it serves as an ideal topic on which to finish the story of understanding the Universe.

The ekpyrotic universe theory was proposed in 2001 by Justin Khoury, Burt Ovrut, Paul Steinhardt, and Neil Turok. In their model, the hot Big Bang was more like a big clap in which two four-dimensional branes collided together. The effect of the collision was to produce the energy, matter, and structure that is now seen in the Universe.

As noted in Chapter 15, M-theory predicts that branes exist in 11 dimensions. Six of these dimensions are curled up too small to be seen but allow strings to vibrate and so give rise to the various fundamental particles. In practice, these six tiny dimensions may be ignored when thinking on cosmological scales. The Universe we perceive, with its three dimensions of space and one of time, accounts for a total of four dimensions. The suggestion of the ekpyrotic universe model is that there is also a fifth dimension (a fourth spatial dimension) that is actually relatively large, and which is not perceived because only gravity can exist there. In fact, the reason gravity is so weak when compared with the other three fundamental interactions may be because some of its effect "leaks out" into this fifth dimension.

In the ekpyrotic universe model, the Universe sits on a visible brane but there is also a nearby hidden brane. The hidden brane is parallel to the Universe but separated at a constant distance from it across the fifth dimension. Particles within the hidden brane interact with particles in the visible brane only through gravity, via the fifth dimension. They may therefore behave like WIMPs. Hiding the non-baryonic dark matter on the hidden brane is an attractive idea and might explain why it is such difficult stuff to locate!

Initially, the visible brane is cold and empty. Then, at some time, the hidden brane shrugs off a lighter brane (like a snake shedding its skin) called a bulk brane. This bulk brane travels across the fifth dimension and collides violently with the visible brane before bouncing off. These branes have mass and so possess kinetic energy. When the branes collide, their enormous kinetic energy is liberated as heat and appears as the hot fireball that has hitherto been thought of as the Big Bang. The enormous temperature creates matter and antimatter particles via pair creation mechanisms, just as described earlier. The collision also triggers the expansion of space within the visible brane.

The colliding branes initially have a flat geometry in the sense described earlier, but there are *ripples* in the bulk brane, caused by random fluctuations at the quantum level. So when the collision happens, it occurs at slightly different times in different places. These ripples in the bulk brane therefore produce variations in the density of the Universe at different places and so ultimately give rise to the departures from uniformity in the CMB radiation that is seen today, and from there grow to become superclusters of galaxies. From here on, the Universe will evolve just as already described in the latter parts of Chapter 16, but the need for inflation has disappeared. The way the four fundamental interactions become separate as the Universe cools, the nucleosynthesis of the light elements, and the production of the CMB, all follow the pattern already outlined.

However, there is one last surprise in store from the ekpyrotic universe model. Between the visible brane and the hidden brane, there is a vacuum, and the vacuum acts like a spring between the branes. Within the Universe, sitting on the visible brane, the effect of the vacuum is to create a repulsive gravitational force that is described by the cosmological constant and attributed to dark energy. Between the branes though, the vacuum force acts as an attraction between them. Today the spring is still being stretched, but eventually it will reach its maximum extension and the branes will start to speed towards each other. Then there will be another collision, another big clap. Notice though that this is not the same as the Big Bang or big crunch scenario that was discussed earlier. In the ekpyrotic universe model, the oscillation only happens in the hidden fifth dimension, between the branes. From the point of view of someone stuck on the visible brane space just keeps expanding, though the expansion keeps getting another push from each successive big clap. It is only from the point of view of the hidden fifth dimension that the Universe appears to be cyclic.

Each "cycle" of the Universe, as it appears on the visible brane, is like that of an open universe. Space continues to expand; stars and galaxies evolve to the ends of their lives and eventually end up consumed by supermassive black holes. The black holes may eventually evaporate into subatomic particles, which then mutually annihilate each other, leading to a dilute sea of photons filling all of space. It is only then, when the Universe has become cold and virtually empty in its heat death, that the vacuum spring between the branes gives rise to another big clap, and the whole process starts again. Matter and radiation are produced from the heat energy liberated from the kinetic energy of the branes, and the whole cosmic evolution of the Universe can happen again. And again. And again. And again. It is a suitably inspiring or humbling thought, depending on how you view it, to realize that "the Universe" may simply be the current version of an infinite cycle of universes stretching forever into the past and into the infinite future.

Book Summary

Congratulations on completing the scientific journey of exploration from quasars to quarks and beyond. In travelling this route, you have seen how understanding the way that the Universe behaves on small scales, and the rules that these smallest constituents obey, provides information about how the Universe behaves on the largest scales, and evolves with time. One of the underlying principles guiding the way on this journey has been the understanding that the *same* laws of physics apply at *all* times and in *all* places throughout the Universe. So, for instance, the principle of the conservation of energy and the behaviour of light that apply on everyday scales are still valid when considering the behaviour of everything from quasars to quarks.

The fundamental constituents of matter have been revealed as particles called leptons and quarks. In particular, the Universe today essentially comprises only first-generation leptons (electrons and electron neutrinos) and first-generation quarks (up and down) which are joined together in triplets called protons (uud) and neutrons (udd), which in turn make up nuclei and atoms. Each particle has an associated anti-matter particle with opposite electric charge and colour charge (where appropriate). These fundamental particles interact with each other by way of four fundamental interactions (known as the electromagnetic, strong, weak, and gravitational interactions) via the exchange of other particles (known as photons, gluons, W and Z bosons, and gravitons).

One of the cornerstones of science is quantum physics, which says that the smallest constituents of matter, such as nuclei, atoms, and molecules, can only exist with specific (quantized) values of energy, known as energy levels. It is the jumps between these energy levels that give rise to the emission and absorption of photons, which are perceived as electromagnetic radiation. The other important feature of quantum physics is that the smallest particles of matter have indeterminate positions and velocities. Bearing this in mind, the most appropriate model of atoms envisages electrons as existing in fuzzy clouds around the central nucleus. Different quantum states of an atom, described by a set of four quantum numbers, correspond to different energy levels and different electron clouds.

The four fundamental interactions are best expressed in terms of quantum theories: quantum electrodynamics describes both electromagnetic and weak interactions, while quantum chromodynamics describes strong interactions. A quantum theory of gravity, however, has yet to be formulated. Because of this, the approximations to reality described by Newton's law of gravity are used to explain most everyday gravitational interactions, and Einstein's general theory of relativity is used to explain gravitational interactions when the masses and accelerations become very large. Vital insight into the way the Universe evolves comes from the understanding that the strength of the four fundamental interactions varies with the energy of interaction. At the highest energies, the four interactions are believed to exist in a single superunified form. As the energy decreases, so the gravitational, strong, weak, and electromagnetic interactions in turn become separated into the four distinct interactions perceived in the Universe today.

This breakdown of unification as energy decreases is the key to understanding the evolution of the Universe. Observations of distant galaxies reveal that their spectra are all redshifted, in accord with the Hubble relationship, implying that the Universe is expanding. Observations of the cosmic microwave background radiation imply that the Universe is cooling. These two observations together show that the Universe began in a hot Big Bang around 14 billion years ago, at which instant, time and space were created. The Universe has cooled and expanded ever since. Particle reactions and decays occurring in the first few minutes of the Universe's history set the scene for the structures observed in the Universe today. Tiny fluctuations in the quantum foam of the early Universe are visible by the imprint they have left on the cosmic microwave background. They also provided the seeds around which galaxies and clusters of galaxies grew in the subsequent millennia.

Within galaxies, stellar systems condensed out of the swirling gas and dust. Stars are the processing factories of the Universe, converting primordial hydrogen and helium into progressively heavier elements through nuclear fusion, before dispersing them throughout space in supernovae explosions, ready to be incorporated into future generations of stars and planets. Planetary systems are now known to be ubiquitous, and the prospect of discovering life elsewhere in the Universe is perhaps closer to reality now than it has ever been.

Homing in on a fairly average spiral galaxy, a fairly average star sits somewhere out in one its spiral arms. Orbiting this star is a small rocky planet, its surface two-thirds covered with water, and with an atmosphere rich in oxygen. On the surface of the planet are many living creatures, including members of one species who are so interested in the origin and complexity of the Universe that they build telescopes and particle accelerators with which to study it. They observe the expansion of the Universe by the redshift of distant galaxies, and the cooling of the Universe by the spectrum of its background radiation. Using particle accelerators they recreate extreme temperatures and examine particle reactions that have not occurred in the Universe for billions of years. These experiments confirm that no epoch or location in the Universe is in any way special, but that at all times and all places the same physical principles hold. Nonetheless, these principles are manifested in a multitude of ways in the Universe we observe around us.

Despite this, the Universe continues to surprise us. Observations of the cosmic microwave background and distant supernovae have revealed that, although the geometry of the Universe is most likely flat and therefore contains a critical density of matter and energy, only around 0.5% of that density comprises the luminous matter in galaxies and stars that can actually be seen. A further 5% of the critical density may be accounted for by baryonic dark matter in the form of MACHOs and the WHIM, and about 26% of the critical density is probably in the form of non-baryonic WIMPs, about whose nature there is virtually no idea. Finally, 69% of the Universe's density is contributed by mysterious dark energy that is accelerating the expansion of the Universe towards its inevitable heat death or a catastrophic big rip.

Acknowledgements

This book has been inspired by various introductory undergraduate science modules which I have contributed to since I joined the staff of the Open University in 1992. I would particularly like to thank Jocelyn Bell Burnell for appointing me to the OU in the first place, and my former colleagues Stuart Freake, Bob Lambourne, and David Broadhurst with whom I worked on the module "S103 *Discovering Science*," where we first explored how to teach the physics of the very small and very large scales in a coherent way. Thanks also to all the OU staff and students over the intervening years who have helped me in understanding the Universe.

Appendix: A Timeline for Understanding the Universe

Year	Event
964	Abd al-Rahman al-Sufi made the first recorded observations of the Andromeda galaxy and the Large Magellanic Cloud.
1054	The supernova that produced the Crab Nebula and pulsar was observed by Chinese astronomers.
1543	Nicolaus Copernicus published his heliocentric theory of the Solar System.
1572	Tycho Brahe made the first detailed observations of a supernova.
1608	Hans Lippershey built the first telescope.
1609	Johannes Kepler published his first two laws of planetary motion.
1610	Galileo Galilei used a telescope to observe phases of Venus and moons of Jupiter, and to discern that the Milky Way is composed of a huge number of faint individual stars.
1619	Johannes Kepler published his third law of planetary motion.
1666	Isaac Newton discovered his law of gravity.
1687	Isaac Newton published his *Mathematical Principles of Natural Philosophy* including three laws of motion and law of gravity.
1750	Thomas Wright correctly speculated that the Galaxy was a rotating body comprising a huge number of stars, held together by gravity.
1755	Immanuel Kant correctly speculated that some nebulae are separate galaxies, calling them "island universes."
1781	Charles Messier published the final version of his catalogue of nebulae.
1783	William Herschel discovered 40 Eridani B, later recognized as the first known white dwarf.
1784	John Goodricke discovered Delta Cephei, the prototype Cepheid variable star.
1785	Charles Augustin de Coulomb proposed the law describing electrostatic force.
1820s	Hans Christian Oersted and André-Marie Ampère showed that electric currents produce magnetic forces.
1830s	Michael Faraday and Joseph Henry demonstrated electromagnetic induction.
1842	Christian Doppler described the Doppler effect.
1848	William Thomson, Lord Kelvin developed the absolute temperature scale.
1862	Alvan Graham Clark discovered the second known, and nearest white dwarf, Sirius B.
1864	James Clerk Maxwell presented his equations unifying electricity and magnetism.
1869	Dmitri Mendeleev laid out the first periodic table of the elements.
1873	James Clerk Maxwell published his theory of electromagnetism.
1890s	Henri Becquerel, Ernest Rutherford, Pierre Curie, and Marie Sklodowska Curie discovered alpha, beta, and gamma radioactivity.
1895	Wilhelm Röntgen discovered X-rays.
1897	J.J. Thomson discovered the electron.
1899	Williamina Fleming and Edward Pickering discovered RR Lyrae stars.

(Continued)

Year	Event
1900	Max Planck introduced the idea of the quantum.
1902	James Jeans proposed a mechanism for interstellar clouds to collapse and form stars.
1905	Albert Einstein published his Special Theory of Relativity, including $E = mc^2$.
1908	Henrietta Swan Leavitt discovered the Cepheid variable period–luminosity relationship.
1910	Ejnar Hertzsprung and Henry Norris Russell independently created their eponymous diagram to illustrate stellar properties.
1911	Ernest Rutherford discovered the atomic nucleus.
1912	Vesto Slipher measured high Doppler shifts for several spiral nebulae.
1913	Niels Bohr suggested an orbital model for electrons in atoms.
1915	Albert Einstein published his General Theory of Relativity.
1916	Karl Schwarzschild derived the formula for the radius of a black hole.
1917	Adriaan van Maanen discovered the first isolated white dwarf.
1918	Emmy Noether showed that conservation laws in physics each result from an associated symmetry of a physical system.
1919	Arthur Eddington observed a solar eclipse to test predictions of General Relativity.
1920s	Erwin Schrödinger and others devised a quantum model of the atom.
1920	Ernest Rutherford identified the proton as a hydrogen nucleus.
1920	Arthur Eddington proposed that nuclear fusion is responsible for powering stars.
1922	William Luyten was the first to use the term white dwarf.
1923	Edwin Hubble observed Cepheid variables in the Andromeda galaxy and calculated its distance, demonstrating it lies beyond the Milky Way.
1923	Louis de Broglie suggested that particles display wave-like behaviour.
1925	Wolfgang Pauli proposed the exclusion principle.
1925	Cecilia Payne deduced that the Sun and other stars are predominantly composed of hydrogen and helium
1927	Werner Heisenberg proposed the uncertainty principle.
1927	George Thomson and Clinton Davisson demonstrated electron diffraction.
1927	Georges Lemaître suggested the Universe began with an "explosion of a primeval atom."
1928	Paul Dirac combined electromagnetism with quantum physics and special relativity and predicted the existence of the positron.
1929	Edwin Hubble discovered the expansion of the Universe.
1930s	Karl Jansky discovered radio waves from beyond the Earth.
1930	Wolfgang Pauli proposed the existence of neutrinos to explain beta decay.
1930	Subrahmanyan Chandrasekhar proposed the upper mass limit for white dwarfs.
1932	James Chadwick discovered the neutron.
1932	Jan Oort proposed that dark matter must exist in the plane of the Galaxy.
1933	Walter Baade and Fritz Zwicky proposed the existence of neutron stars.
1933	Fritz Zwicky proposed that dark matter must exist in galaxy clusters.
1934	George Gamow developed the idea that the Universe began in a singular explosion.
1936	Carl Anderson and Seth Neddermeyer discovered the muon in cosmic rays.
1938	Hans Bethe and Charles Crichtfield proposed the pp chain for hydrogen fusion.
1939	Carl von Weizsäcker and Hans Bethe proposed the CNO cycle for hydrogen fusion.
1939	Robert Oppenheimer and George Volkoff proposed the upper mass limit for neutron stars.

(*Continued*)

Year	Event
1939	Grote Reber identified cosmic sources of radio waves.
1940s	Richard Feynman, Julian Schwinger, and Sin-Itiro Tomonaga developed the theory of quantum electrodynamics (QED).
1943	Carl Seyfert discovered active galaxies.
1948	Ralph Alpher, Hans Bethe, and George Gamow ("αβγ paper") published predictions for primordial nucleosynthesis.
1949	Fred Hoyle coined the term "Big Bang" for the origin of the Universe.
1950	Enrico Fermi presented his paradox between the predicted number of civilizations in the Galaxy, and the lack of any evidence for them.
1956	Clyde Cowan and Frederick Reines confirmed the existence of neutrinos.
1957	Margaret Burbidge, Geoffrey Burbidge, William Fowler, and Fred Hoyle ("B²FH paper") explained the nucleosynthesis of elements in stars.
1960s	Sheldon Glashow, Steven Weinberg, and Abdus Salam developed the theory of electroweak unification.
1960s	Quasars were discovered as quasi-stellar radio sources.
1961	Frank Drake stated his equation for the number of detectable civilizations in the Galaxy.
1962	Murray Gell-Mann coined the term gluon for the carrier of the strong force.
1964	Arno Penzias and Robert Wilson discovered the cosmic microwave background (CMB) radiation.
1964	Murray Gell-Mann and George Zweig proposed the quark model (including up, down, and strange quarks).
1965	Sheldon Glashow and James Bjorken predicted the existence of a fourth flavour of quark, called charm.
1967	Jocelyn Bell discovered the first radio pulsar.
1969	Rashid Sunyaev and Yakov Zeldovich proposed that clusters of galaxies will scatter CMB photons to higher energies.
1970	The first X-ray astronomy satellite, *Uhuru*, was launched.
1970	The first CCD detector was demonstrated.
1973	Harald Fritzsch, Heinrich Leutwyler, and Murray Gell-Mann developed the theory of quantum chromodynamics (QCD).
1973	Brandon Carter proposed the anthropic principle.
1974	Charm quarks were discovered independently at Stanford and Brookhaven laboratories.
1975	Joe Taylor and Russell Hulce discovered the first binary radio pulsar.
1975	Martin Lewis Perl discovered the tau lepton at Stanford Linear Accelerator.
1975	Haim Harari proposed the existence of top and bottom quarks.
1977	The bottom quark was observed at Fermilab.
1978	Vera Rubin and Kent Ford showed that dark matter is needed to explain the rotation curves of galaxies.
1979	Gravitational lensing of a quasar was first observed.
1981	Alan Guth proposed the inflation theory for the early Universe.
1983	W and Z bosons were first observed at CERN.
1986	John Barrow and Frank Tipler published *The Anthropic Cosmological Principle*.
1989	The COBE satellite was launched and measured the temperature of the CMB very precisely.
1995	Michel Mayor and Didier Queloz discovered the first exoplanet.

(Continued)

Year	Event
1995	The top quark was observed at Fermilab.
1998	Saul Perlmutter, Adam Reiss, and Brian Schmidt showed that the expansion of the Universe is accelerating.
1998	Michael Turner coined the term dark energy.
1999	Both the *Chandra X-ray Observatory* and the *XMM-Newton* observatory were launched.
2001	Justin Khoury, Burt Ovrut, Paul Steinhardt, and Neil Turok published the ekpyrotic universe theory.
2001	The WMAP satellite was launched and measured tiny fluctuations in the temperature of the CMB across the sky.
2005	Flavour oscillations were discovered in Solar neutrinos at *Sudbury Neutrino Observatory*.
2009	The *Kepler* satellite was launched to discover transiting exoplanets.
2012	The Higgs boson was discovered at CERN.
2013	The *Gaia* satellite was launched to measure the parallaxes of a billion stars.
2015	The first detection of gravitational waves from merging black holes was made by the LIGO and *Virgo* teams.
2017	The first simultaneous detection of gravitational waves and electromagnetic radiation from merging neutron stars was made.

Index